Water Conflicts in N

Northeast India, apart from being the rainiest in India, is drained by two large river systems of the world – the Brahmaputra and the Barak (Meghna) – both transnational rivers cutting across bordering countries. The region, known for its rich water resources, has been witnessing an increasing number of conflicts related to water in recent years.

This volume documents the multifaceted conflicts and contestations around water in Northeast India, analyses their causes and consequences, and includes expert recommendations. It fills a major gap in the subject by examining wide-ranging issues such as cultural and anthropological dimensions of damming rivers in the Northeast and Eastern Himalayas; seismic surveys, oil extractions, and water conflicts; discontent over water quality and drinking water; floods, river bank erosion, embankments; water policy; transboundary water conflicts; and hydropower development. It also discusses the alleged Chinese efforts to divert the Brahmaputra River.

With its analytical and comprehensive coverage, 18 case studies, and suggested approaches for conflict resolution, this book will be indispensable for scholars and researchers of development studies, governance and public policy, politics and international relations, water resources, environment, geography, climate change, area studies, economics, and sociology. It will also be an important resource for policymakers, bureaucrats, development practitioners, civil society groups, the judiciary, and media.

K. J. Joy is Senior Fellow, Society for Promoting Participative Ecosystem Management, Pune, India.

Partha J. Das is Head of Water, Climate and Hazard Division of Aaranyak, Guwahati, India.

Gorky Chakraborty is Associate Professor, Institute of Development Studies Kolkata, India.

Chandan Mahanta is Professor, Department of Civil Engineering, Indian Institute of Technology, Guwahati, India.

Suhas Paranjape was formerly Senior Fellow, Society for Promoting Participative Ecosystem Management, Pune, India.

Shruti Vispute is a doctoral student at the School of Geography, University of Leeds, UK.

'It is this kind of sound empirical research and careful deliberation on principles that can provide us the solid common ground for obviation or resolution of conflicts over water.'

Mihir Shah, *Distinguished Visiting Professor, Shiv Nadar University, India*

'A must-read for all with an interest in understanding the politics and powers that mark contemporary water interventions and investments.'

Margreet Zwarteveen, *Professor of Water Governance, UNESCO-IHE Institute for Water Education and University of Amsterdam, The Netherlands*

Water Conflicts in Northeast India

Edited by K. J. Joy,
Partha J. Das,
Gorky Chakraborty,
Chandan Mahanta,
Suhas Paranjape and
Shruti Vispute

Routledge
Taylor & Francis Group
LONDON AND NEW YORK

First published 2018
by Routledge

2 Park Square, Milton Park, Abingdon, Oxfordshire OX14 4RN
52 Vanderbilt Avenue, New York, NY 10017

Routledge is an imprint of the Taylor & Francis Group, an informa business

First issued in paperback 2019

British Library Cataloguing-in-Publication Data
A catalogue record for this book is available from the British Library

Library of Congress Cataloging-in-Publication Data
A catalog record has been requested for this book

ISBN: 978-1-138-69725-6 (hbk)
ISBN: 978-0-367-27772-7 (pbk)

Typeset in Sabon
by Apex CoVantage, LLC

Contents

Illustrations

Figures

Tables

Maps

Contributors

Sanjib Baruah is Professor of Political Studies at Bard College, New York, USA, and Honorary Research Professor at Centre for Policy Research, New Delhi, India. He has written widely on Northeast India. His publications include *India Against Itself: Assam and the Politics of Nationality* (1999), *Durable Disorder: Understanding the Politics of North East India* (2005) and the edited volumes *Beyond Counterinsurgency: Breaking the Impasse in North East India* (2011) and *Ethnonationalism in India: A Reader* (2012).

A. C. Bhagabati is a social anthropologist. As a recipient of the Commonwealth Scholarship offered by the Government of New Zealand, he carried out doctoral research among the Maoris in northern New Zealand (1962–65). Professor Bhagabati was Head of the Department of Anthropology at Dibrugarh and Gauhati Universities, Assam, India. He was also Professor and founder Director of the Institute of Social Change and Development, Guwahati. He has published many research papers over the years and his areas of research interest are social change, ethnicity and identity, and tribal transformation process in Northeast India.

Sanchita Boruah is Associate Professor at the Department of Zoology, Dibrugarh Hanumanbax Surajmall Kanoi College, Dibrugarh University, Assam, India, and specialises in Ecohydrology and Hydrobiology of the Brahmaputra River Basin. She has completed her postdoctoral research twice, one at Stirling University, Scotland, and another under the UGC Research Award Scheme and has worked on various national and international research projects.

Gorky Chakraborty is a faculty member at the Institute of Development Studies Kolkata (IDSK), West Bengal, India. He primarily works on development-related issues in Northeast India. As doctoral fellow, he studied the *chars* (river islands) of the river Brahmaputra,

Assam. As a postdoctoral fellow he worked on the Look East Policy and Northeast. He has been a Australia Awards Fellow (2015) at Monash University, Melbourne. He has authored *Assam's Hinterland: Society and Economy in the Char Areas* (2009) and co-authored *The Look East Policy and Northeast India* (2014) and *Accumulation and Dispossession: Communal Land in Northeast India (2017)*.

Runti Choudhury is a doctoral researcher in the Department of Civil Engineering, Indian Institute of Technology-Guwahati, Assam, India. Prior to her research work, she worked extensively on water quality issues in Assam and has been actively involved in several UNICEF-sponsored projects focusing on community water safety and security. She received the Fulbright Nehru Doctoral Research Fellowship for the year 2014–15, where she conducted a part of her research at Columbia University, New York. Her areas of interest include bio-geochemistry, hydrogeology, environmental pollution, water security, safety and sustainability.

Partha J. Das is an environmental researcher at *Aaranyak*, a Guwahati-based, non-government organisation in Assam and heads its 'Water, Climate and Hazard Division'. He completed his doctorate in Environmental Science from Gauhati University, India, specialising in water resources, climatology and hydrology. He has carried out research projects on climate variability in Northeast India, environmental impacts of large river dams in Northeast India, wetland based livelihoods, climate change adaptation, water governance, access to justice and information in water related issues, and community-based forest and wildlife conservation. He has published many research papers, abstracts, reports and articles on these subjects. He is currently working on documentation of climate change adaptation practices, transboundary water governance, community-based flash flood risk management, and water conflicts.

Partha Ganguli is Associate Professor of Economics at Dibrugarh Hanumanbax Surajmall Kanoi College, Dibrugarh, India and currently working on a project supported by CRY. He has worked on various contemporary social and economic issues, especially on aspects of small tea gardens of Assam, which he studied for his doctorate, and the livelihood and demographic patterns of Bodos of Dhekiajuli, Assam, India. His present research interest is water issues, water conflicts and solutions, scarcity, pollution, prospects of tourism in an around water bodies, dams and its socio-economic impacts.

D. C. Goswami is a guest faculty member at the Department of Environmental Science, Gauhati University, Assam, India. He was Colin Mackenzie Chair Professor at Anna University, Chennai in 2004 and was associated with John Hopkins University (USA), Howard University (USA), Berne University (Switzerland), NASA Project (USA), NRSA, Department of Space, India; and Assam Remote Sensing Application Centre at Guwahati. He has specialised in the fields of fluvial geomorphology, environmental science and geoinformatics.

Amelie Huber is a doctoral researcher at the Institute of Environmental Science and Technology (ICTA), Autonomous University of Barcelona, Spain and a member of the European Network of Political Ecology (ENTITLE). Her work focuses on the political ecologies of hydropower development in Sikkim and Arunachal Pradesh, and aspects of conflict, state–society relations and the politics of risk. Her writing has been published in the journals *World Development*, *Capitalism Nature Socialism*, and on ENTITLE Blog (www.entitleblog.org), where she is also a part of the editorial team.

Deepa Joshi is Senior Research Fellow, Centre for Agro-ecology, Water and Resilience at Coventry University, UK. She works on water conflicts in South Asia and is particularly interested in the layering of conflicts within conflicts; political representations in conflicts; and the impacts of dynamic mega-transitions in water on poverty, livelihoods and gender. She has worked extensively on reforms in domestic water policy; the drivers and processes of reforms, and the rhetoric and reality of the focus on equity in evolving policies. She has worked and travelled extensively in South Asia and South East Asia managing programs, conducting policy research and leading local research capacity building. She has also briefly worked in Kenya, Bolivia and Peru and is interested in collaborative, comparative research between South Asia and Latin America, focusing on water reform and water movements in Latin America.

K. J. Joy has been an activist-researcher for more than 30 years and has a special interest in people's institutions for natural resource management both at the grassroots and policy levels. His other areas of interests include drought and drought proofing, participatory irrigation management, river basin management and multi-stakeholder processes, watershed based development, water conflicts and people's movements. He has worked with Bharat Gyan Vigyan Samithi (BGVS) New Delhi, India in its watershed development and resource literacy programme. He was Visiting

Fellow at the Centre for Interdisciplinary Studies in Environment and Development (CISED) Bengaluru, India for a year and was a Fulbright Fellow with the University of California, Berkeley, USA. He has co-authored *Banking on Biomass: A New Strategy for Sustainable Prosperity Based on Renewable Energy and Dispersed Industrialisation* (1997), *Watershed Based Development: A Source Book* (1998) among others. He also co-edited *Water Conflicts in India: A Million Revolts in the Making* (2008).

Nimmi Kurian is Associate Professor at the Centre for Policy Research, New Delhi, India and India Representative, India China Institute, The New School, New York, USA. Her work has focused on the India–China borderlands, comparative sub-regionalism and transboundary resource governance. As a Fellow of the India China Institute, she worked closely with a network of scholars from India, China and The New School on comparative questions of prosperity and inequality in India and China. Kurian is one of the contributors to the India Country Report prepared by the Bangladesh China India Myanmar Economic Corridor (BCIM EC) Joint Study Group, Ministry of External Affairs, Government of India. Her recent publications include *The India China Borderlands: Conversations Beyond the Centre* (2014).

Siddhartha Kumar Lahiri is currently Joint Editor of *South East Asian Journal of Sedimentary Basin Research (SEAJSBR)* published on behalf of the Society of Petroleum Geophysicists (SPG), DU Chapter, India. An observer of contemporary socio-political movements in Assam, he is also Associate Editor of the Assamese magazine *Natun Padatik* (New Foot Soldiers) edited by Professor Hiren Gohain. He completed his doctoral research on 'Basin Evolution, Morphotectonics and Fluvial Processes in the Brahmaputra River System, Assam' from the Civil Engineering Department, Indian Institute of Technology-Kanpur, India and has published in journals such as *Economic and Political Weekly*, *Quaternary International*, *Geomorphology* and *Current Science*.

Jinine Laishramcha is a human rights worker and freelance journalist and presently teaches in the International College at University of Suwon, South Korea. He has completed the Conflict Analysis Certificate Course from United States Institute of Peace, Washington, DC, USA, and attended international humanitarian law training, International Committee of the Red Cross, Geneva. He holds a masters' degree in Human Rights from the Indian Institute of Human Rights, New Delhi, India.

Tseten Lepcha is the founder member of the Affected Citizens of Teesta (ACT) and was General Secretary of the Joint Action Committee of the affected people of Teesta HEP Stage V. He was the Honorary Wild Life Warden of North District of Sikkim, India and member of international plenary board of World Mountain People's Association (WMPA) in France.

Anjana Mahanta is a freelance social scientist and research writer. She writes for leading dailies of the Northeast on the topics pertinent to social and environmental issues. As a guest columnist, she has contributed to *The Telegraph* and *Assam Tribune* on topics ranging from ethnic strife to hydropower dams. She obtained her masters in Economics from Gauhati University, Assam, India and an MPhil in Demography from the Jawaharlal Nehru University, New Delhi, India. She has also worked as a research fellow in the Omeo Kumar Das Institute for Social Change and Development and at Gyan Vigyan Samiti, Assam, India. She received a senior fellowship from the Ministry of Culture, Government of India for a project on the sustainability of performing arts in Assam, India.

Chandan Mahanta is a faculty member at the Indian Institute of Technology-Guwahati, Assam, India and member of the State Water Board and Water Quality Task Force of Assam. He took active role in formulating the State Water Policy for Assam. His current research is focused on sustainability of hydrological resources and ecosystem services, focusing on water quality issues, climate change, sustainable water resource management and environmental management of river basins and wetlands. He has been an ASCE-EWRI visiting fellow at the Utah Water Research Laboratory of the Utah State University, USA. He was awarded a fellowship to undergo the International Programme on Management of Sustainability in the Netherlands. His research team has completed a series of projects towards developing water security strategies and technologies in the Brahmaputra floodplains and also advises concerned departments on these aspects.

Nani Gopal Mahanta is Professor and Head of the Department of Political Science and Coordinator of Peace and Conflict Studies, a UGC innovative programme at Gauhati University, Assam, India. His research interests include issues of peace and conflict resolution, human development and security, insurgency, ethnicity and identity politics. He was selected as the Rotary World Peace Fellow at the University of California, Berkeley, USA from 2002 to 2004.

As a World Peace Fellow, he travelled widely, including Europe and South Asia for his internship and empirical work. He has published articles in various journals, has written a number of books and has often appeared as an expert in media.

Raju Mimi is a news correspondent for *Arunachal Times*, a state weekly published from Itanagar, Arunachal Pradesh, India. He served as Information and Publicity Secretary in All Idu Mishmi Students Union (AIMSU) from 2007 to 2010 and was selected as a spokesperson of the Union from 2010 to 2012. He has previously edited a small local weekly *Veracity* published from Roing from 2007 to 2010. He is a life member of Idu Mishmi Cultural and Literary Society (IMCLS).

Trilochan Pandey is with World Learning/School for International Training (SIT) Jaipur, India. He coordinates the academic and field component of the programme. His areas of interest are development issues in the mountains, social policies and international education in India.

Suhas Paranjape is a graduate from the Indian Institute of Technology-Bombay, Mumbai, Maharashtra, India. He has actively participated in different movements such as People's Science Movement, Adivasi Agricultural Labourers' Movement, etc. He has been engaged as a core team member and consultant in many action research studies and pilot projects undertaken by the Centre for Applied Systems Analysis in Development (CASAD) and Society for Promoting People's Participation in Ecosystem Management (SOPPECOM), Pune, India in the areas of participatory management of natural resources especially in the field of participatory irrigation management. For three years from 1996 to 1999, he worked as a core team member of the Bharat Gyan Vigyan Samiti (BGVS) Odisha, India in its watershed development project across the country. He was a Visiting Fellow at Centre for Interdisciplinary Studies in Environment and Development (CISED) Bengaluru, India for a year. He has co-authored the books *Banking on Biomass: A New Strategy for Sustainable Prosperity Based on Renewable Energy and Dispersed Industrialisation* (1997) and *Watershed Based Development: A Source Book* (1998) among many others. He has co-edited *Water Conflicts in India: A Million Revolts in the Making* (2008).

Chandan Kumar Sharma is Professor at the Department of Sociology, Tezpur University, Assam, India. His areas of research interests include identity politics, social movements, development, environment, migration, agrarian change and culture with special reference to the Northeastern region of India. He has been a Charles Wallace Visiting Fellow to Queen's University, Belfast, UK (2008). He has also been a Visiting Fellow to the Department of Sociology, Delhi School of Economics, New Delhi, and Centre for the Study of Social Systems, Jawaharlal Nehru University, New Delhi, India.

Ghanashyam Sharma is the Program Manager at the Mountain Institute, India. For his doctoral research, he worked on the biogeochemical cycling of nutrients in the traditional agro-forestry systems of the Sikkim Himalayas. During his postdoctoral research on agriculture biodiversity at the United Nations University, Tokyo, Japan, he worked in the South East Asian countries, Yunnan province of China and Northern Japan. He was an associate scholar at the Center for Southeast Asian Studies, Kyoto University, Japan during 2005–07. He was involved in a Global Environment Facility-funded global project called the People, Land and Environment Change of the United Nations University and in Food and Agricultural Organization-Globally Important Agricultural Heritage Systems (GIAHS) initiative on recognition of Sikkim Himalayan Agriculture.

R. K. Ranjan Singh is Director of Academic Staff College, Manipur University, India. His doctorate research is on 'Problem of Land Use in Manipur Valley' from Gauhati University, India. He has carried out extensive research on riverine ecosystems, transboundary water conflicts and climate change, wetlands ecosystems in Northeast. He has played an important role in the popularisation of science and generating consciousness on sustainable development in Manipur, in the establishment of the Manipur State Science and Technology Council, and the Bharat Jan Vigyan Jatha of which he was the State coordinator.

Shruti Vispute is a doctoral researcher at the University of Leeds, UK. Her research arenas explore the phenomenon of water conflicts in India, environmental activism especially around water issues, gender issues in water sector and development politics in the global South. She works as a consultant researcher with the Forum for Policy Dialogue on Water Conflicts in India. She has co-edited the book, *Water Conflicts in India: A Million Revolts in the Making* (2008).

Foreword

Water Conflicts in Northeast India is a compendium of essays on a wide range of water conflicts in Northeast India. Although water-related conflicts have been on the rise in this region in recent times threatening the life and livelihood of people and ecological systems and services they depend upon, not much has been done to systematically engage with these issues academically and intellectually to understand their underlying causes and search for effective ways for prevention, mitigation and resolution. These conflicts arise mainly out of prevailing crisis of water governance resulting from failure of political, administrative, social and economic systems in addressing legitimate water-based needs of the society and the ecosystem. The rationale of the book and its relevance seem to be based on the emerging challenges for water governance in the region marked by its vast untapped water resources potential, unique richness of biological as well as ethnic diversity, active geophysical and hydro-meteorological processes and a human matrix sapped by widespread poverty, inequity and underdevelopment. The scenario gets further complicated due to the transboundary character of most of its river basins spreading across interstate and international boundaries under highly interactive highland–lowland ecosystems vis-à-vis the Eastern Himalayas and the lowland valleys of the Brahmaputra and the Barak rivers.

For the protracted and arduous task of conceptualising, designing, grooming and editing this book, the sole credit should go to Forum (Forum for Policy Dialogue on Water Conflicts in India) – a premier organisation of the country championing the cause of studying and mitigating the problems of water conflicts for more than a decade. However, it is the scholarly contributions of the select group of multidisciplinary academicians and social activists, based on their knowledge and experience of water-related conflicts of the region, that have finally gone into making the book into what it is today. In fact, the book is the outcome of a series of academic and intellectual debates and discussions that the Forum had organised in the Northeast as well as outside on various issues of water conflict in the region. These are reflected in the diversity of issues and their multidisciplinary character covering wide-ranging issues of water-related conflicts.

The book has a total of 17 chapters of fairly extended coverage on a wide range of water conflicts of Northeast India. In the introductory chapter of the book, the editors have done a commendable job in presenting a comprehensive overview and critique of the gamut of issues related to water conflicts in Northeast India that are flagged by different authors in the book. Equally praiseworthy is the meaningful way the seemingly diverse topics representing different chapters are grouped into two internally coherent separate sections, viz., 'fault lines' and 'case studies', which are further subgrouped theme wise. The topics under the 'fault line' section are of broad-based, generic nature with larger spatial coverage, while those in the 'case study' section are of specific type and empirically designed. This carefully planned structural arrangement of the book will no doubt help in providing a fairly systematic and perceptive view of the region's diverse water conflict scenario, besides enhancing the overall thematic integrity of the book.

The major issues of water conflict of the region documented in the book include threats from downstream impact of dams including those of contiguous areas of China and Bhutan and usage of transboundary waters, ineffective measures for management of hazards related to flood, river bank erosion and siltation, health hazards due to fluoride, arsenic and biological contamination of drinking water and seismic survey for oil exploration and hydrocarbon extraction in the Brahmaputra and Barak River Basins. The threat from the downstream impact of hydropower projects is currently the most important cause of water-related conflicts in the Northeast region, especially Assam, creating great apprehension and deep sense of insecurity in the public mind, often leading to massive rallies and loud protests. A number of case studies and scholarly articles in the book highlight the contours and complexities of some of these issues in the light of their scientific, socio-economic and environmental contexts.

Let me conclude by saying without reservation that the book, more precisely the themes it espouses, has the topical relevance and intellectual precision needed to raise the level of awareness and understanding of its readers on Northeast India's most prized, yet poorly governed, water resources and its least understood, highly fractured water conflict scenario.

D. C. Goswami

Acknowledgements

This book, *Water Conflicts in Northeast India*, is a reworked version of 'Water Conflicts in Northeast India: A Compendium of Case Studies' brought out as an internal publication by the Forum for Policy Dialogue on Water Conflicts in India (Forum) in 2013. While engaging with the different types of water conflicts unfolding in the Northeast India, especially through this process of documentation of case studies and the couple of workshops we held to discuss the case studies we realised the complete de-link between what is happening in the Northeast, especially the conflicts around hydropower development, and the rest of the country and the world. We thought a formal publication would help in taking these conflicts out of the Northeast and make them more visible among a larger readership. This is the motivation in bringing this out as a book.

It has been a long-drawn-out process. The process of identifying the potential authors, deciding on the case studies and the actual documentation began in 2010. All the case studies have been peer reviewed. The Forum organised a national workshop, 'Water Conflicts in the North East: Issues, Cases and Way Forward', at Guwahati in December 2010 that brought together all the case study writers and a few other experts to discuss the case studies and provide critical feedback. The authors finalised the case studies taking the peer review comments and the inputs from the workshop into account. One more round of revision took place after receiving the comments and suggestions received from the anonymous reviewer from Routledge. Thus, the authors have engaged with the case studies for years. We are deeply grateful to all the authors for their contribution and also their perseverance and patience in responding to our persistent queries over a long period of time.

The critical remarks and suggestions from the peer reviewers greatly helped in improving the case studies. We sincerely thank the reviewers for the kind cooperation, inspirational support and valuable inputs they provided in spite of their busy schedules. Also, we are thankful to the anonymous reviewer for the critical comments and suggestions as they helped in restructuring the book and also in revising some of the case studies and also the introductory chapter.

We thank Aaranyak for their collaboration in organising the workshops and commissioning the case studies. We also thank the participants of the above-mentioned national workshop who provided important feedback on the case studies.

We specially thank Prof. D. C. Goswami and Prof. A.C. Bhagabati, both Members, Advisory Committee of the Forum, for their guidance. They have been a true inspiration behind the Forum's work in Northeast India.

We are grateful to Suchita Jain for working on the maps included in the book and other team members of the Society for Promoting Participative Ecosystem Management (SOPPECOM)/Forum for their help in bringing out this book. We also acknowledge the efforts put in by various copy editors and proofreaders at various stages of this publication.

Arghyam Trust, Bangalore, has been supporting the work of the Forum since 2008 and the original compendium was brought out as part of this support. We would like to acknowledge the financial support and encouragement provided by Arghyam and special thanks to Amrtha K. of Arghyam for her enthusiastic support and inputs.

Last but not least, our thanks to Routledge for agreeing to publish this under its banner and bringing it out in such an elegant manner. Shoma Choudhury, Commissioning Manager of Routledge, has been a source of support and encouragement. Many a times we faulted on the deadlines set by us to submit the manuscript. Without her perseverance and patience, we would have long given up on this. Forum's relationship with Routledge dates back to 2008 when it brought out the first book of the Forum, *Water Conflicts in India: A Million Revolts in the Making*.

As mentioned earlier, the motivation for bringing out this book is to link the different types of water-related conflicts lately emerging in the Northeast with an interested audience from the rest of India and the world. We sincerely hope that this book provides the 'starting block' for a more sensitive – socially and environmentally – engagement with these conflicts and the issues they bring forth by the civil society, academia and the political class.

K. J. Joy, Partha J. Das, Gorky Chakraborty,
Chandan Mahanta, Suhas Paranjape
and Shruti Vispute

Abbreviations

AASU	All Assam Students' Union
ACT	Affected Citizens of Teesta
ADB	Asian Development Bank
AGP	Asom Gana Parishad
AIMSU	All Idu Mishmi Students Union
AJYCP	Asom Jatiyatabadi Yuva Chatra Parishad
AOC	Assam Oil Company
BIS	Bureau of Indian Standards
BOC	Burmah Oil Company
BRB	Brahmaputra River Basin
BRO	Border Roads Organisation
CDM	Clean Development Mechanism
CEA	Central Electricity Authority
CISMHE	Centre for Interdisciplinary Studies of Mountain & Hill Environment
CRPF	Central Reserve Police Force
CWC	Central Water Commission
CWPRS	Central Power and Water Research Station
DACC	Dam Affected Citizens Committee
DADC	Downstream Anti Dam Committee
DBCC	Dibang River (Talon) Basin Consultative Committee
DBWC	Dibang Basin Welfare Committee
DRT	Dibrugarh-Rongagora-Tinsukia
EIA	environmental impact assessment
EJ	environmental justice
GDP	gross domestic product
GIS	geographic information system
GLOFs	Glacial Lake Outburst Floods

GOI	Government of India
GOS	Government of Sikkim
GSDP	gross state domestic product
GU	Gauhati University
HEP	hydroelectric project
IMCLS	Idu Mishmi Cultural and Literary Society
JAC	Joint Action Committee
JOM	Jiadhal Nadi Baan Pratirodh Oikya Mancha
KBR	Kangchendzonga Biosphere Reserve
KMSS	Krishak Mukti Sangram Samiti
LSEG	Lower Subansiri Expert Group
MLA	Member of Legislative Assembly
MoEF	Ministry of Environment and Forests
MoU	memorandum of understanding
MoWR	Ministry of Water Resources
MP	Member of Parliament
MPCB	Manipur Pollution Control Board
MW	megawatt
MWR	Ministry of Water Resources
NASBO	National Sikkimese Bhutia Organisation
NASS	Nagarik Sangathan Sikkim
NEAA	National Environmental Appellate Authority
NEC	North Eastern Council
NECWBD	Northeast Commission on Water and Big Dams
NEEPCO	North Eastern Electric Power Corporation
NEWRA	North East Water Resources Authority
NGOs	nongovernmental organisations
NHPC	National Hydro Power Corporation
NWAC	National Water Advisory Committee
NWP	National Water Policy
OIL	Oil India Limited
PCBA	Pollution Control Board of Assam
PHED	Public Health Engineering Department
PRIs	Panchayati Raj Institutions
PWSS	Piped Water Supply Schemes
R&R	resettlement and rehabilitation
R-o-R	run-of-the-river
SIBLAC	Sikkimese Bhutia Lepcha Apex Committee
SOPPECOM	Society for Promoting Participative Ecosystem Management
SPDC	Sikkim Power Development Corporation

SWP	State Water Policy
THD	Tipaimukh High Dam
UNEP	United Nations Environment Programme
WCD	World Commission on Dams
WHO	World Health Organization
WRD	Water Resources Department

1 Understanding water conflicts in Northeast India

K. J. Joy, Partha J. Das, Gorky Chakraborty, Chandan Mahanta, Suhas Paranjape and Shruti Vispute

Luitote mor ghar, Luitei mor par
Luitei je bhange, garhe, xopon moromor

(*Luit** is my home, yet it is a stranger to me,
It is the *Luit* that makes and mars my dreams of love)
. . . .

Bistirna paarore
Axonkhya jonore
Hahakar xuniu
Nixobde nirobe
Burha luit tumi
Burha luit buwa kiyo?
Noitikotar skhalan dekhiu
Manobotar patan dekhiu
Nirlajja alax bhawe buwa kiyo?

You have heard the cries of millions on your vast shores,
Yet O' Old *Luit* why do you flow in silence?
You have seen morality eroding and humanity collapsing,
Yet why do you flow shamelessly and sluggishly?

(By Bhupen Hazarika; Translated from Assamese by Patha J. Das)
*Luit is the colloquial name of the Brahmaputra River in Assam and Hazarika often refers to Brahmaputra as Luit

There is now a growing awareness about water conflicts in India and the world over. However, most of it has been of the doomsayer variety so far, given the regularity with which it is predicted that the next world war will be over water! This growing anxiety is not matched by systematic work on understanding water conflicts. While individual cases have been studied fairly intensely – for example, studies on the conflict around the Sardar Sarovar dam on the Narmada River would comprise a bibliography running into several pages – there is very little methodical and sustained effort at documenting water conflicts in India, particularly those unfolding in relatively remote parts of the country like the Northeast.

The Forum for Policy Dialogue on Water Conflicts in India (the Forum) has been engaged in a sustained initiative of documenting several kinds of water conflicts in India as one of its regular research and documentation programmes. This book, *Water Conflicts in Northeast India*, is part of this continued initiative.[1]

We have organised this introductory chapter to the book around four sections. The first section profiles the Northeast region in administrative, ethnic, bio-physical, sociocultural and economic terms, providing a broader context to the water conflicts unfolding in the region. Section two discusses the nature of water conflicts in Northeast India. Section three is about the book itself in terms of the processes, structure and the case studies. The Introduction concludes with the fourth section, which provides certain pointers to engaging with the water, or more specifically hydropower, related conflicts in the region as a possible way forward.

The Northeast region

Northeast India refers geopolitically to a group of eight States –often referred to as seven sisters and one brother – located in the Northeastern corner of the country. The region includes Arunachal Pradesh, Assam, Manipur, Meghalaya, Mizoram, Nagaland, Tripura and Sikkim.[2] The region covers a total area of 262,185 sq. km, and is located between latitudes 21.57°N and 29.30°N, and longitudes 88°E and 97.30°E (Map 1.1).

According to the 2011 census, the total population of India's Northeast is 45,587,982, which is 3.77 per cent of India's total population. It has the lowest density of 174 persons per sq. km. The 2011 census also showed that the vast majority of the region's population – about 82 per cent – lives in rural areas.[3]

Rich ethnic diversity

Nearly one-fourth of the population in the Northeast is tribal. About 145 tribal groups live here, representing one of the world's richest ethnic diversity. Tribals constitute the majority in the States of Arunachal Pradesh, Meghalaya, Mizoram, Nagaland, Manipur and Sikkim. These tribes mainly belong to the Indo-Mongoloid groups. Many of these tribal groups are very small in terms of population – nearly half of them have less than 5,000 people. The Government of

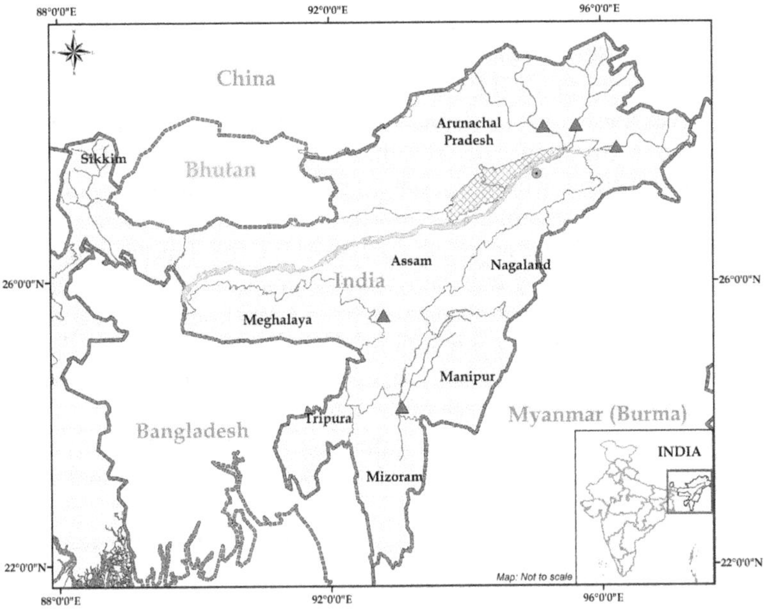

Map 1.1 Map of Northeast India
Source: SOPPECOM, Pune

India has recognised 121 tribal groups as Scheduled Tribes – 12 in Arunachal Pradesh, 5 in Nagaland, 29 in Manipur, 17 in Meghalaya, 14 in Mizoram, 19 in Tripura, 23 in Assam and 2 in Sikkim.[4] The main tribal groups of the region are the Naga,[5] Mizo, Lushai, Hmar, Kuki, Chin, Bodo, Dimasa, Karbi, Kachari, Borok, Tripuri, Reang, Jamatia, Garo, Jaintia, Adi, Aka, Apatani, Nyishi, Monpa and Paite (Fernandes *et al.* 2008). Anthropologists have recognised about 64 ethnic groups of the region as major tribes. Populationwise, the Nagas (living in Nagaland, Arunachal Pradesh, Manipur and Assam) are the largest tribal group, followed by Bodos, Khasi-Jaintia, Garo, Mizo, Tripuri, Karbi, Kachari, Kuki and Chakma. This region represents a sort of 'ethnological transition zone' between India and the neighbouring countries of China (Tibet), Burma and Bangladesh (Ali and Das 2003). Apparently, 220 languages are spoken in the region belonging to three language families – Indo-Aryan, Sino-Tibetan and Austric.[6] The tribals of the region speak different languages belonging to the Sino-Tibetan linguistic family except for the Khasi group, who speak a dialect of Mon-Khmer linguistic group, which is a part of the Austro-Asiatic languages.

Land of abundant natural resources

This region is part of both the Himalaya (the Eastern Himalaya) and the Indo-Burma biodiversity hotspot – one of the 35 biodiversity hotspots in the world. Geo-ecologically, it is a part of the Eastern Himalayas known for its richness in water resources, biodiversity and ethnic and cultural diversity as well. The region has about 164,000 sq. km of forest cover, which is about 60 per cent of its total geographical area. It accounts for about one-fourth of India's forest cover.

Settled agriculture is mainly practised in the Brahmaputra and Barak valleys. Shifting cultivation is still practised in the hills of the region. An estimate suggests that about 443,300 households earn their livelihoods from shifting cultivation.[7] The reported area under shifting cultivation is about 22,480 sq. km, which is approximately 10 per cent of the total geographical area of the region.[8]

The region is drained by two large river systems of the world – the Brahmaputra and the Barak (Meghna). Both are transnational rivers cutting across bordering countries. It is also one of the wettest regions of India. As a result, the region is endowed with abundant water resources and hydropower potential. As natural resources are owned by communities in most States in the region, States' effort to gain control over these resources has remained a source of disgruntlement for the people. Moreover, indigenous communities are discontented because natural resources are being utilised for development without involving local people and traditional institutions in the decision-making process. Such a development paradigm, coupled with disregard for traditional institutions and community opinion, has prepared the ground for conflicts. The nature of water-related conflicts in the region is typical of its socio-cultural complexity and political sensitivity.

Nature of water conflicts in the Northeast

Though the Northeastern region harbours colossal water resources, the ongoing efforts to harness its vast hydropower potential through a series of hydropower projects have posed an unprecedented threat to the water and ecological security of the region, thereby leading to food insecurity as well. Hydropower dams involve the setting up of large infrastructure, which in turn leads to deforestation and disruption of forest ecosystems and reduction of biodiversity. The indigenous people living near dam sites who are largely dependent on forests and rivers for livelihoods are feeling threatened. Further, the widespread detrimental impacts on the downstream flood plains, the river regime, aquatic biodiversity, groundwater domain, wetlands and consequent effects on agriculture and environment can lead to loss of livelihoods as well as both in and out migration, thus increasing the possibility of conflicts and social unrest.

The hydropower potential of the region has attracted national and international attention, with the result that more than 168 hydropower projects with large river dams are being planned for the region. A large number of memorandum of understanding (MoU) have been signed with various companies, prompting the then Minister of State for Power at the Centre Mr Jairam Ramesh to make a cryptic comment in 2008 that the Northeast is hit by an MoU virus! Many of these projects are in different stages of execution by public and private sector companies. There is widespread concern over the observed and probable social and environmental impacts of these hydropower projects in the region. Protests against the detrimental downstream impacts of large dams in Assam have assumed the proportions of a mass movement. The State has reacted, most often, violently to the peaceful means of the anti-large dam movement in the region. The most recent example is the killing of two anti-dam activists in Tawang of Arunachal Pradesh in police firing on 2 May 2016.[9]

Flood, riverbank erosion and sand casting are three serious water-induced hazards that have significantly affected people's lives, livelihoods, agriculture and economy of States like Assam. Floods are also disastrous for Tripura and Manipur. The State's approach to flood management has left a lot to be desired. Adoption of short-term measures like embankments as the main method of flood containment, lack of proper and socially acceptable resettlement and rehabilitation (R&R) package and consistent failure to protect riparian areas from collapsing into rivers tell a story of poor governance and management of these water-induced disasters. People are not only unhappy with inadequate rehabilitation and relief, they have also started protesting against inappropriate structural interventions and the financial corruption of vested interest groups in the government in perpetuating ad hoc and ineffective flood mitigation infrastructure.

The quality of drinking water is another area of growing concern in the region. Government actions have proved to be ineffective to counter the increasing contamination of groundwater with fluoride and arsenic resulting in serious health hazards. According to a study done back in 2003–04 by the North Eastern Regional Institute of Water and Land Management, Tezpur (Assam), the concentration of arsenic in groundwater exceeded the permissible level (50 mg/L by World Health Organization) in parts of Assam (20 districts out of 24 districts), Tripura (3 districts out of 4 districts), Arunachal Pradesh (6 districts out of 13 districts), Nagaland (2 districts out of 8 districts) and Manipur (1 district out of 9 districts) (Singh 2004).

Transboundary issues like the building of dams by China or its alleged attempts to divert the Brahmaputra River within China are now topics of hot debate in the region. The upstream–downstream linkages within the region and the contiguous Himalayan areas are also engendering other types of conflicts. Landslides induced dams getting breached or diffused in Bhutan and Tibet have caused catastrophic floods in downstream areas in Arunachal Pradesh and Assam. The sudden release of water to rivers from upstream dams within the region, especially in Bhutan, has caused devastating flash floods. Lack of coordination between countries, as well as States, sharing the river basins is a major obstacle in resolving these problems in the Northeast region.

Resource-based development often has significant negative transboundary and even global costs, ranging from cross-border damage in the case of upstream hydropower dams, to the regional and global impacts of deforestation (Cronin 2009). Transboundary conflicts, both transnational and interstate, emerge when States at the upstream of a water resource use the water available to them to extract more power, and when States at the downstream use other forms of power such as military power to get more water (Yumnam 2009). The issue of large-scale hydropower development in the Eastern Himalaya has already caused a simmering tension among China, India and Bangladesh. Similarly, India's talk of interlinking of rivers to transfer water from the 'surplus' basins to the 'deficit' ones has also created anxiety and tensions among the neighbouring countries like Nepal and Bangladesh. Stress on the socio-economic and geopolitical systems occurs when changes in water demand due to rapid population growth, land use shifts or development of technology outpace institutional capacity (McNally *et al.* 2009).

The perceived threat felt by Bangladesh due to India's Tipaimukh dam in Manipur and China's alleged plan to divert the Brahmaputra elucidate the potential of transboundary conflicts over the use of water resource. This can naturally put the residents of this region under heightened threats, both of water hazards and lack of water security. There is also an apprehension, which could be rather remote, in the region that even after the projects are completed, these dams can remain potential targets for a missile or bomb attack by enemies or insurgent groups. Residents strongly believe that the region is a seismically high-risk prone one and can cause devastation in future similar to a high-magnitude earthquake hitting the region in the past. The earthquakes of 1897 and 1950, both of Richter magnitude 8.7, are among the most severe in recorded history and they caused extensive landslides and rock falls on hill slopes, subsidence and fissuring in the valley and changes in the course and configuration of the Brahmaputra and its tributaries (Goswami and Das 2003) and thereby significantly changed the geomorphologic and hydrologic regimes of the region. Thus, a multiplicity of issues related to water conflicts have led to an increased sense of insecurity and anger among local communities.

The book: structure and cases

It is in recognition of the need to understand the nature of water-related conflicts of the Northeast region that the Forum embarked on a new initiative to document selected water conflicts in the region. This initiative and its output is expected to raise awareness among national and international readership about the water conflict issues of the region and their sociocultural and environmental contexts, leading to an informed discussion and debate. It is premised that a proper understanding of the issues involved would also help in their resolution and prevention. It could also be immensely useful for policy makers and advocacy organisations to develop strategies and instruments for resolving these conflicts, which is a crying need to ensure sustainable development through peace building in this strife-torn and politically volatile region.

The book, including the Introduction, has 17 chapters. They cover a wide range of issues and types of conflicts including oil exploration in the riverbeds, water quality, drinking water, riverbank erosion and embankments, water policy, dams and hydropower projects and trans-boundary issues. Since hydropower projects have emerged as the main source of water conflicts in the region, it is no wonder that a large number of the case studies in the book are related to hydropower development in the region (see Map 1.2 for the locations of the case studies).

The chapters following this Introduction are divided into two parts: fault lines and case studies. The chapters under Part I are not case studies of specific water conflicts; instead they are thematically more generic in nature (as in the case of Chapter 2 by Bhagabati or Chapter 8 by Baruah); all of them cover a large geographical area like a State or the entire Northeast region itself (see Chapter 5 by Huber and Joshi); or they are emerging situations and areas for conflicts or the conflicts are latent (Chapter 3 by Choudhury, Mahanta and Mahanta is an example of this). The second and larger section, Part 2 Case studies, contains chapters which are cases of specific water conflicts in Northeast India.

Fault lines

Part I opens with Bhagabati's chapter, 'Damming of rivers and anthropological research: an introductory note'. He had presented this as a paper in a seminar at Gauhati University way back in 1983. We reproduce this paper with his kind permission, because we find that even after more than 30 years, the issues that it brings forth are still very relevant. Bhagabati argues for the need to incorporate anthropological research inputs in development planning, especially relevant in the context of hydropower projects in the region. If this is not done, the author warns us, we will have to pay a heavy socio-ecological cost.

One of the important emerging fault lines in the waterscape of Northeast in general, and Assam in particular, is the water quality issue. Though till now no significant conflict around water quality has manifested in the region, Choudhury, Mahanta and Mahanta's chapter, 'Water quality in Assam: challenges, discontentment and conflict', highlights the latent conflicts around it. This is a theme that has not been discussed as much as other types of water conflicts. The chapter shows that contamination of water by arsenic, fluoride and heavy metals, along with bacteriological contamination, has raised widespread concerns regarding drinking water security and safety in the State of Assam, leading to resentment and mistrust, particularly towards government bodies entrusted with maintaining water quality. The authors are of the opinion that water contamination has an implicit role in conflicts related to water security and safety in Assam.

Sikkim, like Arunachal Pradesh, has been in the forefront of developing hydropower in the last one decade. In Chapter 4, Sharma and Pandey describe how the government's strategy of increasing the State's revenue through hydropower has put a huge stress on the local environment, the people and their culture. Most of the hydropower projects are coming up in areas that are sacred, and spiritually and culturally important for indigenous communities, especially the Lepchas. They also take note of the seismic occurrences in the already unpredictable Himalayan ecosystems, call for revisiting the concept that hydropower projects are 'green, clean and safe'. They concede that hydro projects can bring considerable income to the State and its people, but only if they take into account the bio-physical, sociocultural and economic environment of the region. They also suggest that Sikkim should declare a few of its rivers as the 'cultural heritage and sanctuary' for coming generations.

In the next chapter, 'Hydropower conflicts in Sikkim: recognising the power of citizen initiatives for socio-environmental justice', Huber and Joshi show that the new hydropower development discourse is couched in ostensible win–win scenarios: securing energy for the rapidly developing national economy; accelerating development in hitherto 'backward' but hydro-potent areas; and generating 'clean' energy that cut through the earlier dam-related critiques. Taking a political ecology approach, they try to show how the entire environmental/water governance gets 'depoliticised' by transferring environmental governance to the State or State-backed private technological–managerial control in the specific context of a process of eroding democracy and an 'imposed . . . benevolent despotism' in Sikkim. They recommend an urgent review of the official dam-related policies and regulatory frameworks, and suggest means to strengthen emergent socio-environmental justice initiatives.

An emerging issue leading to polarised discourse is privatisation and commercialisation of water, especially after India embarked on the economic reform path. Sharma's chapter, 'State Water Policy of Assam 2007: conflict over commercialising water', tells the story of how the newly formulated Draft State Water Policy (SWP) of 2007, with its pro-commercialisation thrust, has become a point of contestation in Assam. Sharma argues that the same policy has been continued by the post-colonial state, and the draft SWP 2007 continues to reflect the colonial legacy of converting commons to private property. Civil society objected to the draft SWP mainly because it tries to bring in commodification of water by facilitating private participation, provides for river interlinking and construction of big dams, excludes civil society and bows to the union government's pressure on the State governments to fall in line with the National Water Policy (NWP). Some of the positive outcomes of the mobilisation of civil society around the draft SWP include the formulation and submission of an alternative water policy draft, setting up of a Task Force to review the draft SWP, holding a number of review meetings with civil society representatives and inclusion of the position of the community as the primary repository of rights to water. After this process, the SWP has been finalised and handed over to the Government of Assam in late June 2009. However, it has not been promulgated as policy yet.

The last two chapters in this section, namely, 'Water conflicts in Northeast India: the need for a multi-track mechanism' by Mahanta and 'Whose river is it, anyway? The political economy of hydropower in the Eastern Himalayas' by Baruah, not only take an overarching view of the issues but also point to possible ways to engage with them. Mahanta warns that neglect of the important issues associated with large dams and hydropower projects – cultural displacement, alienation and demographic change; displacement; militarisation; violation of the right to development and human security; and protection of indigenous rights – could lead to further escalation of conflicts in the region. He argues for the democratisation of the water sector and multi-track institutional arrangements to provide space for negotiated settlements or exploring 'win–win' solutions accommodating opinions of all stakeholders. He proposes a five-point agenda by way of an alternative approach to water resource development for the Northeast region: (1) acceptance of democratic principles by all statutory bodies while pursuing the development agenda; (2) institutional mechanisms that provide participation of all stakeholders including local people from Panchayati Raj Institution (PRIs) and upwards in the environmental impact assessment (EIA) reports, public hearings and decision making; (3) acceptance of the World Commission on Dams' (WCD) recommendations about the construction of large dams; (4) recognition of the right to water as a fundamental right; and (5) constitution of a Northeast Commission on Water and Big Dams as a regulatory and adjudicatory body.

Baruah's chapter comprehensively sums up the discourse on hydropower development in the Northeast. Though Baruah's chapter draws largely on the Lower Subansiri project and the controversies surrounding it, it brings into focus the political economy of hydropower development in the region and engages with almost all issues that the various case studies in the book later bring out. He makes a fundamental difference

> between the hydropower projects of postmillennial India and the multipurpose river valley projects of an earlier period in India's post-colonial history. In the mid-twentieth century large multipurpose river valley projects were taken up, to develop a river basin region. They were driven by the spirit of decolonisation itself. . . . By contrast, what is being designed and built these days are almost all single-purpose hydropower dams with power to be produced and sold for a profit by private as well as public sector companies.
> (Baruah, Chapter 8, this volume)

This difference in the nature of the projects and the increasing realisation that these projects may not contribute substantially to the development of the local people and its economy is an important factor in the hydropower-related conflicts in the region.

Case studies

In this section on specific water-related conflict cases, we begin with conflicts centred on flooding and erosion processes caused by the rivers in Assam. There are three case studies on this issue: the first two deal with flood and riverbank erosion and the government's approach to control these disasters through structural means like embankments, which has been a bone of contention in the country; the last one deals with issues related to the control over the new land, called *char* lands, created through the process of erosion and deposition in the braided channels of the Brahmaputra River.

In Chapter 9, Das discusses the increasing discontent of the people in the Jiadhal Basin area in the Dhemaji District of Assam because of the increasing devastation caused by floods and erosion year after year, government's apathy towards proper maintenance of the embankments, rejection of community's opinion on matters related to flood and erosion protection and the wrong siting of a newly planned embankment. This discontent led to fierce resistance by the people, and the government adopting violent means to control it. The case study brings out how flawed governance and poor flood mitigation can trigger and escalate conflicts.

Chapter 10 by Lahiri brings out a different dimension of riverbank erosion in the Brahmaputra Basin. He discusses how the particular bio-physical features of the Brahmaputra coupled with recurring earthquakes in the region have resulted in the erosion of both the bank-lines, thus leading to the expansion of the channel-belt within a relatively short period of 90 years. The case study location, Rohmoria (in Dibrugarh District of Assam), is the worst affected, especially in terms of loss of land. Though the situation has been extremely hazardous, most of the State government's plans to check bank erosion had not been implemented until the discontentment of the people spilled over onto the streets, including the prolonged oil blockade that started in August 1999.

The third chapter on riverbank erosion, aptly titled 'The *char* dwellers of Assam: flowing river, floating people' by Chakraborty deals with the unique situation of the people who live in the mid-channel bars or islands of the Brahmaputra and its tributaries locally known as *chars*. These mid-channel bars or islands are formed under the conditions of floods as the restless river erodes them in one place while adding and extending them in another, completely inundating one in one place, raising an entirely new one in another. The life of the *char* land dwellers is lived within this complex shifting reality and is characterised in turn by settled life on a *char*, forced migration and a possible return to the old patch of land if and when the Brahmaputra returns it. Chakraborty describes in detail how the floods create and affect the floating population of *char* land dwellers and the numerous kinds of conflicts that it leads to. The issues get further compounded because of the historical prejudices against the dwellers of the *char* lands who are mostly Muslims and are often described as 'illegal immigrants' from Bangladesh.

The next set of case studies is around the search for and extraction of oil in the Brahmaputra and Barak Basins. The two case studies, 'Seismic survey for oil exploration in the Brahmaputra River Basin, Assam: scientific understanding and people's perceptions' by Boruah and Ganguli and 'Hydrocarbon extraction in Manipur and its impact on Barak downstream' by Laishramcha deal with contestations over river-bed oil exploration and extraction. Boruah and Ganguli make the point that though extraction and production of oil is beneficial to the society in diverse ways like providing direct and indirect employment, on the flip side, there is a lack of transparency in the concerned authorities regarding the technologies used for seismic surveys and oil exploration. Their likely negative impacts on the aquatic ecosystems and water-dependent livelihoods of people coupled with uneven sharing of costs and benefits have resulted in differing perceptions and contestations over oil exploration and extraction along the Brahmaputra River in Assam.

Laishramcha's study describes the impending oil extraction in Manipur and how the adverse impact of the water contamination due to oil extraction activities will make it even more difficult for people to earn their livelihoods in the entire region around the Barak River from its source in Manipur through Assam and Bangladesh. This situation, according to the author, will engender severe conflict over the ever-shrinking land and water resources among the ethnic and religious communities.

The block of three case studies that follow is devoted to conflicts around hydropower projects which, as mentioned earlier, is the most important water conflict issue in the region. Between them, they cover a wide range of issues around hydropower including displacement, loss of livelihoods, various types of ecological impacts especially in the downstream, seismicity and the fragile nature of the Himalayas, flooding, ethnicity-culture-sacred landscapes, faulty environmental impact assessments and so on. Of course, there are many, many more hydropower-related conflicts in the region and the effort here has been to pick a few to bring out the different dimensions of hydropower-related conflicts.

Mimi's case study, 'The Dibang Multipurpose Project: resistance of the Idu Mishmi', brings out the two main contentions around the project in the Dibang Basin in Arunachal Pradesh: one, the justification of the project on grounds of economic viability as the displacement is considered to be negligible; and two, the fears of the Idu Mishmi community, the primary inhabitants in the region, that the influx of outsiders into the region because of dam building will lead to a demographic imbalance in the Dibang Basin. For example, the 17 planned projects in the Dibang Basin together would bring in about 100,000 outsiders, whereas the total population of the Idu Mishmi community is only about 11,000. The hydropower onslaught on the Dibang, revered by the Idu Mishmis, has spelt 'cultural genocide' and stimulated a debate about the right to ancestral land, identity and culture. The youth, students and literary people, organised by the All Idu Mishmi Students Union (AIMSU) and the Idu Mishmi Cultural and Literary Society (IMCLS), are in the forefront of the anti-dam movement. The State and local politicians have used the police and paramilitary forces to break the people's resistance. These excesses have left a deep impact on the psyche of the local Idu Mishmi community. The fear of intensification of State repression probably has led the leadership of the movement recently to reconsider their anti-dam stand and start negotiations with the government on a revenue sharing model between the state and the local community.

Ranjan Singh, in Chapter 15, discusses the case of the Tipaimukh High Dam (THD) on the border of Manipur and Mizoram. Originally the THD was designed to contain flood waters in the lower Barak Valley, but later a hydropower component was added to it. Singh brings out the issues of displacement and deprivation of livelihoods especially of a large number of indigenous communities, mostly belonging to the Zeliangrong and the Hmar people. The THD is also a trans-boundary issue, as the people of Bangladesh are worried about the possible changes in the flow pattern of the Barak (Meghna) River downstream of the dam after its construction.

During the last two decades or so the Northeast has witnessed many popular resistances against the hydropower projects, and Sikkim has been in the forefront of this. Lepcha's chapter, 'Hydropower projects on the Teesta River: movement against mega dams in Sikkim', captures the historic struggle led by Affected Citizens of Teesta (ACT) against the dams on the Teesta in Sikkim. Lepcha shows how numerous projects adversely affect the ecosystem, livelihoods, religion, cultural identity and political rights of the people and violate the sacred landscape which has been an important cause of discontent. The heavy influx of outside workers for the projects seems to be changing the demography of the place, as these workers stay on in the State, affecting the social, economic and political situation, and also exerting great pressure on its sparse resources including land. Lepcha narrates the story of the movement of the affected people who came together under the banner of ACT, and the historic hunger strike of 915 days that forced the government of Sikkim to invite the ACT for negotiations.

The final case study deals with the geopolitical positioning of the region and the transboundary nature of the river systems which together have given rise to many water conflicts across national boundaries. In Chapter 17 Kurian discusses the much talked about 'Chinese angle' to the Northeast water conflicts. She examines China's water resource choices within the context of its overall water policy directions, the possible conditions under which China is likely to exercise these choices, the ripple effects they are likely to have across the borders and some key concerns that have flown downstream. She argues that only a borderlands perspective can help redraw the conceptual

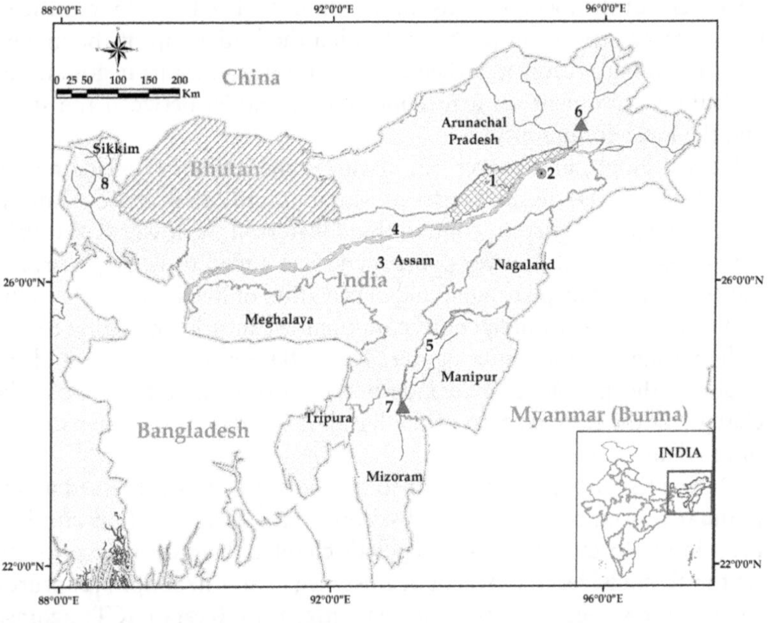

Map 1.2 Map of the Northeast showing the locations of the cases

Source: SOPPECOM, Pune

1. Conflicts over embankments on the Jiadhal River in Dhemaji District, Assam
2. Riverbank erosion in Rohmoria: impact, conflict and people's struggle
3. The char dwellers of Assam: flowing river, floating people
4. Seismic survey for oil exploration in the Brahmaputra River Basin, Assam: scientific understanding and people's perceptions
5. Hydrocarbon extraction in Manipur and its impact on Barak downstream
6. The Dibang Multipurpose Project: resistance of the Idu Mishmi
7. Tipaimukh High Dam on the Barak River: conflicting land and people
8. Hydropower projects on the Teesta River: movement against mega dams in Sikkim
9. An uneven flow? Navigating downstream concerns over China's water policy (not shown in the map)

toolkit to look at trans-border resource governance challenges that a shared neighbourhood brings. This would depend, according to her, on the willingness of China, India and other countries in the region to begin a conversation on public goods and common pool resources in the region and their sustainable and equitable management.

Possible way ahead

The Northeast India is a unique example of a region with plentiful water where different types of intense water conflicts are emerging. The experience in the region also shows that 'surplus' water resources alone do not prevent conflicts, if not supported by required institutional mechanisms for their sustainable development, equitable distribution and access, safeguards against water hazards and democratic governance. So far, the riparian States have failed to prevent water conflicts, which could not have been expected as the region is endowed with colossal water resources.

In this concluding section, we do not present a blueprint of how to resolve the different types of conflicts which constitute the subject matter of this book. That is neither feasible nor desirable. Instead, we discuss a few tentative approaches to the various issues these conflicts bring forth, so that they become more tractable.

As the chapters included in the book indicate, there are different types of water-related conflicts unfolding in the Northeast region – water hazards (floods, erosion and landslides), water quality, river bed oil exploration and extraction, transboundary water issues and hydropower development. However, the most significant conflict in the present times concerns what Sanjib Baruah calls 'the Great Leap Forward in hydropower generation' (Baruah 2012; also, Chapter 8, in this volume) in the region.

Looking at water conflicts from an environmental justice framework

The different types of water-related conflicts unfolding in the region have one thing in common: the issue of (in)justice, both environmental and social. Very often water conflicts are seen from the lens of 'security'[10] and there is an urgent need to go beyond the security-centred conceptualisation of conflicts. In fact all the case studies in this volume bring out that all conflicts, whether they concern river bank erosion, oil exploration or hydropower projects, are all related to the issue of justice and fairness – who gains and who loses, how equitably the costs and benefits are shared, who decides and how democratic the decision-making process really is. These are also core issues of water governance. Looking at water conflicts from an environmental justice (EJ) framework would help us to move away from dominant techno-management approach to water governance and forefront the political economy and political ecology questions underlying the different types of water conflicts.

EJ, mainly seen as a label for global movements against privatisation of water and other natural resources, dams and displacements, unequal sharing of costs and benefits and so on and so forth, originated in the early 1980s in the context 'distribution of environmental "burden"', especially disposal of toxic waste with race as the focus. Today it encompasses a large terrain and can tie together issues of both environmental and social justice. It talks of re-allocation of income, resources and power. The trivalent conception of EJ – distribution, recognition and participation – can help us move away from the increasing depoliticisation of the discourse (techno-management and efficiency fixes, tendency to universalise and naturalise the water discourse), a tendency that is gaining ground in all sectors with economic reforms since the 1990s. Instead the need is to foreground the political economy/ecology questions of conflicts, contestations, rights, justice, equity, sustainability and participation/democratisation around water governance in general and conflicts in particular which EJ can do. This could also provide us with a different set of agenda around different types of water-related conflicts – in terms of research, action and policy engagement.[11]

Main drivers of hydropower development in the region

As mentioned earlier, presently the conflicts around hydropower projects seem to overshadow all other water conflicts in the region. Hence we shall try to dwell a little more into these conflicts.

With the publication of the Report of the WCD in 2000, along with heightened civil society action against large dams, especially in the 1990s, there was a belief that the days of mega dams were over. However, within a decade, mega dams seem to be back with a bang. The massive hydropower development plans in the Northeast, often described as the future powerhouse of India, became an integral part of this solitary power-centric engagement in water resources development. This was quite in line with the vision and strategy of the policy makers of India to fast track hydropower development in the country. A vision document circulated by the Central Electricity Authority of India (CEA) in 2001 provided preliminary ranking studies of about 400 hydropower dams with a total potential of about 107,000 MW. In 2003, the then Prime Minister Atal Bihari Vajpayee announced the 50,000 MW initiative that included 'prefeasibility reports' on 162 new projects with an aggregate capacity of 47,930 MW (ADB 2007: 13; as cited in Baruah 2012). These projects were to be completed by 2017, and were to be followed by another drive to add at least 67,000 MW additional hydropower capacity in the subsequent 10-year period (International Rivers 2008: 7; as cited in Baruah 2012, also, Chapter 8, in this volume). According to one estimate, in a 10-year period, Arunachal Pradesh alone proposes to add hydropower capacity which 'is only a little less than the total hydropower capacity added in the whole country in 60 years of Independence' (Human Rights Law Network 2008: 3; as cited in Baruah 2012, also, Chapter 8, in this volume). 'This projected pace of dam-building and the scale of India's hydropower development plans is unprecedented – nothing short of an attempt at a Great Leap Forward in hydropower generation' (Baruah 2012; also, Chapter 8, in this volume). The hydropower initiative in the Northeast needs to be placed in the backdrop of this grand design of hydropower development in the country as 'a significant part of India's untapped hydropower potential is located in the rivers of the Eastern Himalayas' (Baruah 2012, also, Chapter 8, in this volume). India's plan to expand its power generation potential – both hydropower and thermal power – is to keep up its high growth trajectory through industrialisation and urbanisation. Thus, the main driver of such large-scale hydropower generation in the Northeast is arguably not the development of the region as such, but the export of hydropower to the rest of India to fuel its high-growth economy. It amounts to what some civil society activists call 'resource colonisation' of the Northeast by the rest of India. This also partially explains why single-purpose hydropower projects are being put on a fast track, and not the multipurpose projects which were taken up in the post-independent era. These multipurpose projects could have met some of the critical developmental needs of the particular region, especially if these projects were planned as part of integrated river basin development, including overdue components like flood moderation, drinking water needs, irrigation and navigation.

For the region, it is revenue generation for the State governments that is driving the hydropower development plan. The National Hydro Power Corporation (NHPC) generally provides 12 per cent of the power to the State government as royalty. In the case of certain private companies, they even pay 15 per cent as royalty, as in the case of the Lower Demwe project on the Lohit River by Athena Energy Ventures.[12] To get a rough idea of how much revenue the government would earn, let us consider the example of Arunachal Pradesh, which would have the largest hydropower generation if all planned projects come true.

> It has a projected hydropower potential of around 57,000 MW. As of July 2011, the state government has signed MoU with power developers for 147 hydropower projects of over 40,000 MW installed capacity. The state government hopes to earn around Rs. 12,000 crores a year through the export of power to various parts of the country once the full capacity is developed. The government believes that hydropower development will catalyse the overall development in this Himalayan state.
>
> (Vagholikar 2013)

Profit for the private sector is the other motive, as most projects are going to be implemented by private parties. In Arunachal Pradesh, of the 132 projects with a total capacity of about 40,140 MW, for which the government has signed MoU with developers, 120 are with private companies (Vagholikar and Das 2010). The main attraction for the private developers is the 'free' fuel; unlike thermal and other sources of power, in the case of hydropower plants, though huge investments are required initially, once they are built, the operational costs are minimal (Baruah 2012; also, Chapter 8, in this volume). It is this prospect of windfall profits that has pushed the private sector to come in a big way, including companies that have been unheard of in the power sector (Deka 2010: 4; as cited in Baruah 2012, also, Chapter 8, in this volume).

Climate change is the latest motivation to push the hydropower agenda, as production of hydropower is seen as low-carbon energy and thus part of the climate mitigation strategy. It is also seen as 'green' energy, which makes it eligible for funding under the Clean Development Mechanism (CDM), in return for carbon credits (ADB 2007). More than one-fourth of the projects applying for funding under the CDM are hydropower projects, and most are located in the Himalayas of India and China (Dharmadhikary 2008). The claim that hydropower is green power has always been contested.[13] This is all the more true in the context of the Eastern Himalayas, as it is one of the climate change hotspots in the world, and given its fragile nature, the large-scale damming of the area may exacerbate impacts of climate change even further. Large dams invariably enhance people's vulnerability and risk in a changing climate. Given the scientific evidences of large dams contributing to global warming as well as becoming potential sources of risk due to climate change, it has become imperative to do proper risk assessment in a holistic sense for each of the proposed or ongoing projects and for every river basin in which multiple projects are being planned in the Northeast (Vagholikar and Das 2010).

The 'run of the river' nature of most of these projects is also being used by the pro-hydropower lobby to give a clean chit to the projects in terms of their environmental as well as social impacts, as the destructive submergence can be much less compared to the conventional, behind-the-dam storage-centric hydropower production. However, many argue that these projects in the Northeast may not adhere to the definition of run of the river projects, as they 'involve large dams, which divert the river waters through long tunnels, before the water is dropped back into the river at a downstream location after passing through a powerhouse. These projects are promoted as being "environmentally benign" as they involve smaller submergences and lesser regulation of water as compared to conventional storage dams' (Vagholikar and Das 2010). Ghanashyam Sharma and Trilochan Pandey in their case study, 'Harnessing energy potential in fragile landscapes: Exploration of conflicts and emerging issues around hydropower developments in Sikkim', aptly describe what extensive tunnelling has done to the popular psyche in Sikkim. They say:

> The power projects have already earned the sobriquest of *durey musa* (mountain beaver), the local people say *Darjeeling le Gorkhaland hosh is na-hosh, tara Sikkim chain todkaland bhayo* (whether Darjeeling gets Gorkhaland or not, but Sikkim has become a holeland, i.e. the land of tunnels).
>
> (Sharma and Pandey, Chapter 4, this volume)

Finally, we have the Chinese angle to hydropower development in the Northeast. China's South to North Water Diversion project that aims to take waters from the Yangtze River and divert it to the Yellow River has created apprehensions among its neighbouring countries. India is particularly concerned about China's plan to divert the Yarlung Tsangpo (Brahmaputra). As Nimmi Kurian states in her chapter in this book:

> As part of its response strategy, India is seeking to establish a case for user rights on the Brahmaputra by tapping the hydropower potential of Arunachal Pradesh. The moves to expedite plans to develop mega storage hydroelectric projects on the sub-basins of the Siang, Lohit and Subansiri rivers can be read as India's moves to institutionalise such a norm. But this raises a more fundamental question: what is the likelihood of India and China subscribing

> to the norm of prior use as a possible basis for equitable water sharing, given that both have chosen not to ratify the only existing international treaty governing shared freshwater resources. One also needs to look beyond formal and legal instruments to 'soft law' and an entire range of innovative, informal processes that allow for flexibility and consequently greater measure of success.
> (Kurian, Chapter 17, this volume)

This seems to be the best bet for India, given the asymmetries in the geo-politics of the region.

Where do we go from here?

First, the term 'resource' has been interpreted variously, for example, that which can be drawn upon as a means of help or support; natural advantages, especially of a country, as forest, oil deposits; available wealth or property; skill or ingenuity in meeting any situation; any way or method of coping with a difficult situation, etc. There are other interpretations and classifications of resources but in each and every interpretation great emphasis is laid upon the situational aspect which determines the necessity for using a particular resource. In other words, it can be argued that the use of a particular resource is determined by its historical necessity, and normatively speaking the necessity should emerge from within the 'people' under whose jurisdiction the resource is available. In an indigenous society like the Northeast governed by customary laws and institutions, resource is always treated as people's resource where its use and allocation is decided by the people through their traditional institutions. A lack of proper understanding of these aspects of resource and their utilisation in an ethnic *milieu* like the Northeast will lead to a hiatus between the people and the State which is bound to generate conflict (Chakraborty 2005). The contemporary situation related to dam building in the region is an indication in this direction.

Second, what has been the case with the colonial construct of 'wasteland' during the British period seems to be replicated with hydropower in contemporary Northeast. In the early colonial period following the Lockeian logic any land that did not yield revenue to the Crown was labelled as 'waste'. This included the community lands too. Land which is not assigned by the sovereign belongs to the sovereign (*res nullius*) that included no man's land (*terra nullius*). These lands which were termed as 'wastelands' were to be used for revenue generation by the colonial state. But in the ethno-history of the 'people' who practised usufruct cultivation land left fallow were to be used by the community in future. Today the historical fallout of this colonial construct termed 'wasteland' and their utilisation in Assam in particular and Northeast in general is evident to all (Chakraborty 2012). And under such a situation, when the State is bent on defining the property rights in the name of 'hydropower development' through private property regime in water it is bound to create conflict. Our case studies in the volume indicate a similar situation.

Thus, we need to discard the dominant idea that water flowing out to the sea is a waste. This mindset, which is still prevalent in the country, led to a water management strategy centred on dams. This is being reproduced in the Northeast as well. The lesson that our hydrocracy refuses to learn is that water is a resource embedded within ecosystems performing specific ecological functions and delivering explicit ecological services on which the well-being of natural and human systems is critically dependent; we cannot treat it as a freely manipulable resource detaching it from the socio-ecological niches and contexts. For example, too many of our mega projects, whether big dams, hydropower projects, diversions or interlinking schemes, treat it as freely manipulable and thus harm the long-term viability and sustainability of the resource itself. Our wetlands and rivers are already in bad shape. It is time we stop looking at these water systems simply as resources to be mined. Otherwise, we will end up spending more in managing conflicts, compared to what we get from our projects! (Joy *et al.* 2008).

We would urge the concerned authorities to put a stop to all project-related activities, in whatever stage they are, till a comprehensive review of all projects is carried out, taking into account the various objections raised by academics, activists and affected people, and made available in the public domain. The manner in which the EIAs of various projects have been carried out has attracted serious criticism. This issue has been sharply brought into focus by several case studies in the book.[14] Issues range from conflict of interest to credibility of the people (and the institutions) who carry out EIAs, their lack of independence and at times capability, lack of transparency, a denial of the right of project affected people to be involved in decision making, project planning and implementation, and stage-managed public hearings. Often the EIAs appear *fait accompli*, and apparently not a single project in India has been rejected based upon the EIA report. The issue with EIAs is not limited to the Northeast alone; this is the norm all over the country. It is high time that the entire process is streamlined and restructured in a more transparent, democratic, participatory, scientific and objective manner. The recent measures implemented by the

Ministry of Environment and Forests and Climate Change to accredit qualified EIA consultants following stringent selection procedures is welcome but not adequate. The need for cumulative impact assessments is gaining ground as well, because a number of projects are coming up in the same river basin. We cannot elaborate here on how an objective EIA should be conducted due to space constraints. There has been enough civil society engagement[15] on this issue, which could provide the basis for determining the methodology of objective EIAs. The Ministry of Environment and Forests (MoEF) should demonstrate the political will to streamline processes, establish the required protocol and order fresh EIAs for all projects in the Northeast as part of this comprehensive review.

Social Impact Assessments (SIAs) are as important as EIAs especially in a sensitive region like the Northeast known for its ethnic diversity and socioculturally vulnerable communities. Several case studies in the book have brought out the intrinsic vulnerabilities of different ethnic communities in the region in terms of their livelihoods, culture, customs and traditions as well as their sacred spaces. An in-depth socio-anthropological assessment[16] of the likely impacts of the hydropower projects on the lives and livelihoods of ethnic people should be included in the comprehensive review.

Hydropower projects are being promoted in the Northeast region often for exogenous reasons. It is time to reverse this trend. The starting point of this comprehensive review process should be the developmental needs of the region, taking into account its bio-physical, socio-economic and cultural diversity. The region should be able to define its developmental agenda and use its natural resources with people's participation. India as a nation also needs to rethink its high growth developmental pathway, which is putting enormous stress on natural resources. In the age of climate change, the need and wisdom of switching over to a developmental pathway with a low energy and water footprint cannot be emphasised enough. This would help to redraw the future energy policy and strategy of the country. Demonstrated measures could be taken to bridge the gap, for instance, energy efficiency measures, and tapping renewables like wind, solar, biomass and small hydro, thus opening up a wide range of options. This would help to move away from a large hydro-centric energy strategy, and as a result, the need for large-scale hydropower production in the Northeast could be substantially reduced.

It is the responsibility of the concerned governmental agencies and departments, with the participation of academic institutions and the civil society, to articulate different alternatives and present these before the people so that they can make informed choices from a basket of options. This is quite in sync with some of the key recommendations of the World Commission on Dams: one, the least cost (environmental, economic, social and cultural) option; and two, the people who would be affected should have a decisive say in this choice (WCD 2000). If a certain number of hydropower projects get chosen from the region with the consent of its people to provide power to the rest of India, then it should be on the condition that the revenue accruing from these projects should be shared fairly with the affected communities, and that too with minimal ecological and social costs.[17]

There is clearly an institutional vacuum in the region with regard to hydropower development. Though multiple projects are being planned in each of the river basins and sub-basins, there is not a single river basin institution in place. In fact what is urgently needed is a system of nested institutions that start from the micro level, may be a village or micro-watershed, and proceed upwards to the basin level board or authority (Joy and Paranjape 2009). The institutional framework should be guided by the principle of subsidiary in the sense that the decisions that can be taken by a lower level/scale institution should not be controlled or usurped by a higher level/scale one. The institutions at appropriate levels/scales should be able to make choices among different options and also negotiate the settlement of disputes.

One would earnestly urge all those who are concerned with hydropower development and other potentially conflict-generating water infrastructure projects in the Northeast to stop all such contentious projects, in whatever stage they are, till the above steps are completed. If this is not done, the writing on the wall clearly points to the inevitability of further escalation of water conflicts in the region. Though water conflicts need not always be bad or detrimental (and also sometimes needed), we must realise that they do have serious implications for both the society and the ecosystems and we need to take account of them. In the Northeast, the poorest of the poor are under threat, especially the small ethnic groups, as well as the sources of their water – the rivers, springs, wetlands – and their very survival.

Notes

1 For details of the Forum's work, please see http://waterconflictforum.org, http://conflicts.indiawaterportal.org.
2 Sikkim was a hereditary monarchy till 1975, when it merged with India to become the 22nd state of the country. Sikkim is called the 'brother' and the other seven states are the 'sisters'.
3 www.censusindia.gov.in/2011-prov-results/paper2-vol2/census2011_paper2.html, Provisional Population Totals Paper 2, Volume 2 of 2011 (India & States/UTs), Accessed on 28 February 2013.

4 http://censusindia.gov.in/Tables_Published/SCST/ST%20Lists.pdf, Accessed on 13 February 2013.
5 The Naga is a generic name of a group of tribes such as Ao, Sema, Konyak, Angami, Lotha, Chakesang, Khiemungam, Phom, Sangtam, Yimchungre, Zeliang and Rengma (Jaswal 1998).
6 www.iitg.ernet.in/rcilts/phaseI/n_e.html, Accessed on 16 January 2013.
7 http://dspace.nehu.ac.in/bitstream/1/7716/3/Land%20use%20(SK%20 Barik).pdf, Accessed on 16 January 2013.
8 http://dspace.nehu.ac.in/bitstream/1/7716/3/Land%20use%20(SK%20 Barik).pdf, Accessed on 16 January 2013.
9 http://indianexpress.com/article/india/india-news-india/monk-among-two-killed-in-tawang-police-firing-2781542/.
10 The security school looks at conflicts in terms of various insecurities like physical, military, economic, environmental/ecological, societal and political (Source: 'Conflict Analysis', notes prepared and circulated by Lambrecht Wessels during the one-day training programme on conflict analysis organised as part of the CCMCC (Conflict and Cooperation in the Management of Climate Change) research project partners' meeting at Hague, the Netherlands, in April 2016).
11 See Joy *et al.* (2014) where an attempt has been made to apply EJ in the context of water and water reallocations and emerging conflicts and contestations.
12 Minutes of the 24th Meeting of the Standing Committee of National Board for Wildlife, p. 18, 13 December 2011, New Delhi (http://moef.nic.in/downloads/public-information/mom-24-13.12.11.pdf), as cited in Baruah (2012).
13 For example see Rudd *et al.* (1993), Lima *et al.* (2007), International Rivers (2008), DelSontro *et al.* (2010).
14 This is also detailed out in Vagholikar (2013).
15 For example, the South Asia office of International Rivers has sent a set of recommendations to amend the Environment Impact Assessment (EIA) notification, signed by 32 prominent persons from the country, to the MoEF on 8 January 2013. The recommendations are based on over six years of experience of the EIA notification, and the adverse comments and observations the EIA process has received from judicial and quasi-judicial bodies, as well as Committees and Panels, including those set up by Courts and the Ministry of Environment and Forests (MoEF). For details of the recommendations write to Samir Mehta, the South Asia Coordinator of International Rivers, at samir@internationalrivers.org.
16 Prof. Bhagabati's chapter in this volume, 'Damming of Rivers and Anthropological Research: An Introductory Note' can provide important pointers for such assessments.
17 In Nepal in the case of hydropower projects, 15 per cent of the revenue generated is shared with the affected people through the elected bodies for development purposes.

References

ADB. 2007. *Hydropower Development in India: A Sector Assessment.* Report by K. Ramanathan and P. Abeygunawardena. Manila: Asian Development Bank.

Ali, A. N. M. I. and I. Das. 2003. Tribal situation in North East India. *Stud: Tribes Tribals*, 1(2): 141–148.

Baruah, S. 2012. Whose river is it, anyway? The political economy of hydropower in the Eastern Himalayas. *Economic and Political Weekly*, 47(29), 21 July 2012: 41–52.

Chakraborty, G. 2005. 'Structural Changes and Resource-Industry Linkages in Mizoram: Some Issue to Ponder'. In: Das, G. (ed.). *Structural Change and Strategy of Development: Resource-Industry Linkages in North-East India*. New Delhi: Akansha Publishing House, pp. 154–158.

Chakraborty, G. 2012. *Roots and Ramifications of a Colonial 'Construct': The Wastelands in Assam*. Occasional Paper 39. Kolkata: IDSK.

Cronin, R. 2009. Mekong dams and the perils of peace. *Survival: Global Politics and Strategy*, 51(6): 147–160.

Deka, Kaushik. 2010. 'Dams: The Larger Picture'. In: Talukdar, Mrinal and Kishor Kumar Kalita (eds.). *Big Dams and Assam*. Guwahati: Nanda Talukdar Foundation, pp. 2–9.

DelSontro, T. *et al.* 2010. Extreme methane emissions from a Swiss hydropower reservoir: Contribution from bubbling sediments: *Environmental Science & Technology*, 44(7): 2419–2425.

Dharmadhikary, S. 2008. *Mountains of Concrete: Dam Building in the Himalayas*. Berkeley, CA: International Rivers.

Fernandes, W., G. Bharali and V. Kezo. 2008. *The UN Indigenous Decade in Northeast India*. Guwahati: North Eastern Social Research Centre.

Goswami, D. C. and P. J. Das. 2003. The Brahmaputra River, India: The eco-hydrological context of water use in one of world's most unique river systems. *Ecologist Asia*, Special Issue: Large Dams in Northeast India – Rivers, Forests, People and Power, 11(1), Jan–Mar 2003: 9–14.

Human Rights Law Network. 2008. *Independent People's Tribunal on Dams in Arunachal Pradesh: Interim Report*. New Delhi: Human Rights Law Network.

International Rivers. 2008. *Dirty Hydro: Dams and Green House Gas Emissions, Fact Sheet of International Rivers*. www.internationalrivers.org/files/dirtyhydro_factsheet_lorez.pdf (Accessed on 18 May 2010).

Jaswal, I. J. S. 1998. 'Tribes of Northeast India: Ethnic and Population Aspects'. In: Karotemprel, S. (ed.). *The Tribes of Northeast India*. Shillong: Centre for Indigenous Cultures, pp. 22–29.

Joy, K. J. and S. Paranjape. 2009. 'Water Use: Legal and Institutional Framework'. In: Iyer, R. (ed.). *Water and the Laws in India*. New Delhi: Sage.

Joy, K. J., Biksham Gujja, Suhas Paranjape, Vinod Goud and Shruti Vispute. 2008. 'A Million Revolts in the Making: Understanding Water Conflicts in India'. In: Joy, K. J., Biksham Gujja, Suhas Paranjape, Vinod Goud and Shruti Vispute (eds.). *Water Conflicts in India: A Million Revolts in the Making*. London, New York and New Delhi: Routledge, pp. xvii–xxxii.

Joy, K. J. *et al.* 2014. Re-politicising water governance: Exploring water re-allocations in terms of justice. *Local Environment: The International Journal of Justice and Sustainability*, 19(9). Special Issue: Water Rights, Conflicts, and Justice in South Asia, pp. 954–973.

Lima, I.B.T. *et al.* 2007. Methane emissions from large dams as renewable energy resources: A Developing nation perspective. *Mitigation and Adaptation Strategies for Global Change*, 13(2): 193–206.

McNally, A., D. Magee and A. T. Wolf. 2009. Hydropower and sustainability: Resilience and vulnerability in China's powersheds. *Journal of Environmental Management*, 90(3): 286–293.

Rudd J.W.M. *et al.* 1993. Are hydroelectric reservoirs significant sources of greenhouse gases? *Ambio*, 22: 246–248.

Singh A. K. 2004. *Arsenic Contamination in Groundwater of North Eastern India.* Proceedings of National Seminar on Hydrology with focal theme on 'Water Quality' held at National Institute of Hydrology, Roorkee, 22–23 November 2004.

Vagholikar, N. 2013. 'Demwe Lower Hydroelectric Project in Lohit River Basin: Green Clearances Bypass Ecological and Socio-Cultural Concerns'. In: Das, Partha J. *et al.* (eds.). *Water Conflicts in Northeast India: A Compendium of Case Studies*. Pune: Forum for Policy Dialogue on Water Conflicts in India, pp. 126–133.

Vagholikar, N. and P. J. Das. 2010. *Damming Northeast India.* Guwahati: Action-Aid.

World Commission on Dams. 2000. *Dams and Development: A New Framework for Decision-Making* [The Report of the World Commission of Dams]. London: Earthscan.

Yumnam, A. 2009. *Political Economy in the North East: Contextual Issues, Presidential Address*. 11th Conference on Northeastern Economic Association held at Indian Institute of Technology Guwahati, 18 December 2009.

Part I

Fault lines

2 Damming of rivers and anthropological research

An introductory note[1]

A. C. Bhagabati

The fact that major development programmes in the spheres of land and other natural resources have important socio-ecological implications is widely recognised these days. This recognition has generated research, bearing upon practical problems of land and natural resources exploitation and management. In India, it is only during the last few years that socio-ecological problems have come to be perceived as matters deserving serious attention. Since 1981, the Planning Commission and, subsequently, the newly created central Department of Environment have taken some steps to initiate coordinated action research programmes in various aspects of man and ecology. Non-governmental organisations such as the Centre for Science and Environment have also been established in the past few years to take stock of the situation and stimulate ecologically oriented research.

As far as the discipline of anthropology is concerned, the recognition of the interlinkage between man and environment is nothing really new. Man, as a part of the ecosystem achieving distinctive types of adaptation and also a manipulator of his environment with the aid of his culture have been major premises of anthropological studies for a long time. It is therefore only natural to expect the involvement of anthropology in research bearing upon emergent socio-ecological problems. After all, anthropology in its applied aspect is concerned not merely with fact-finding inquiries but also with giving advice and active participation in the planning processes of the national governments.

There is, however, little or no attempt in our country to incorporate anthropological research inputs in development planning. Major gaps may thus remain in our planning process for which the price to be paid in terms of socio-ecological consequences later on will be quite heavy. It is possible to identify a large number of emergent areas where specialised anthropological research has practical relevance. To make my point, I shall take up only one example which has significance for Northeast India. It is the proposals now being actively pursued by organisations of the government such as the Brahmaputra Board, the North Eastern Council (NEC) and the North Eastern Electric Power Corporation (NEEPCO) of damming different rivers in the Brahmaputra Basin. Damming rivers is one type of developmental activity that has important socio-ecological implications.

Before discussing the Brahmaputra proposals, let us consider in a general way the nature of linkage between high dams and anthropological research with examples from other parts of the world. This will help us in identifying and anticipating issues that are relevant in our own region.

A dam, by definition, is a structure built across a stream or river to hold back water in the form of a reservoir or lake which may then be used for a number of purposes: generation of hydroelectric power, reduction of peak discharge of flood water, irrigation and so on. As a basis for economic development, the multipurpose dam holds special importance in underdeveloped countries. At the same time, this type of development project has important social and ecological impacts in the terrain around the dam site. Building of a major dam usually sited in hilly regions generates tremendous pressures on the traditional inhabitants of the locality. The floral and faunal assemblages of the area may also suffer adversely if proper care is not taken for habitat management as a part and parcel of the planning and construction of the dam. As for the inhabitants, they are required to cope with diverse and frequently drastic changes requiring utmost capacity for adjustment.

The Kariba dam in Africa may be taken as an example. Built between 1956 and 1960, this 420-feet-high dam is located at Kariba Gorge across the Zambezi River, on the border between Zambia and Zimbabwe. The dam made possible the generation of 8,000 MW of electricity, most of which goes to Zambia. But the dam also submerged a 2,000 sq. miles area creating the vast Lake Kariba. The inundated land was part of the traditional homeland of the agriculturist Ba Tonga tribesmen. Some 57,000 tribesmen had to be uprooted and resettled in lands provided for them in sparsely populated areas around the new lake and in other parts. Thousands of wild animals had to be evacuated in an operation that came to be called 'Operation Noah'. Though I have not been able to get hold of the copies, I remember seeing a series of publications called the Kariba studies published in the early 1960s and edited by the anthropologist Cora Dubois. The series, I believe, is more in the nature of salvage anthropology concerned with the investigation and documentation of modes of life about to be disrupted by a forced change of habitat.

Another example is provided by the hydroelectric dam of Itaipu built across the Parana River which forms the frontier between Brazil and Paraguay in South America. The dam is a joint undertaking of these two countries. Said to be a dam of massive proportions, as high as a 62-storey building and capable of generating 12,600 MW of electricity by 1988 when all its 18 turbines were installed, its social and ecological costs have been huge. Hundreds of farmers had to be evicted from millions of acres of arable land to be inundated. A tribe of Indians, the Avagurani, was forced to move to higher barren land. Thousands of wild animals had to be rescued from the rapidly rising waters of the lake caused by the dam. To my knowledge, no study to help effective rehabilitation or to reduce relocation shocks in an uncongenial habitat of the affected aboriginal populations was initiated by any of the governments.

The last example I wish to refer to is the well-known Aswan High Dam of Egypt, begun in 1959 and completed in 1970. While this 365-feet-high dam yields enormous benefit for Egypt as a whole by providing over 2,000 MW of electricity, irrigating fields and controlling the Nile flood for the first time in history, it also meant the submergence of some famous archaeological sites and the traditional homeland of the Nubian people in both Egypt and Sudan. Nubia, the land that once stretched along the Nile from Aswan in Egypt to Merowe in Northern Sudan, now lies under the vast man-made lake that is 400 miles long and on an average five miles wide. Some 100,000 Nubians in Egypt and Sudan were displaced by the submergence. They were relocated on recently reclaimed land of Kom Ombo in Egypt and Khasim al Quirbah in Sudan. As Hussein Fahim, an Egyptian anthropologist points out (Fahim 1981: 87), only one third of the Sudanese Nubian population was displaced, whereas the entire Nubian community within the Egyptian territory, numbering about 48,000, had to leave their homeland. The new settlement at Kom Ombo which has come to be known as 'New Nubia' is designed as a compact area of 40 villages for the three ethno-linguistic Nubian groups who before relocation each lived in a separate region along the Nile.

The Aswan High Dam Project is somewhat unique among projects of this kind in the sense that since 1960 when plans were announced concerning displacement of the Nubian, local and foreign anthropologists undertook many surveys, studies and investigations among the Nubians, including compilation of comprehensive record of Nubian culture before its exposure to change. But even here, as Fahim points out, there was little effective research bearing upon the likely impact of imposed and rushed relocation of such a large, essentially peasant population, who also constituted a distinctive cultural minority in the Egyptian context. There was little policy-oriented research which should have come as an integral part of the planning process of the high dam itself. The likely socio-economic implication of the project ought to have been thoroughly researched well ahead of time as a component of the project planning itself. Since this was not done, many of the subsequent developments on the sociocultural and economic fronts among the relocatees were not anticipated. Relocation was a far from successful programme. For instance, the basically agricultural economy of the Nubians suffered a setback; there was a rise in social tension in the relationship between the relocatees and the dominant non-Nubian Saidiys who made up the host community; the traditional language showed signs of decay, especially among the youngsters. In sum, the Nubians in Egypt are said to depict a general state of dissatisfaction as the 'New Nubia' could not provide a viable economic base or a landscape for settling down and feeling at home.

These examples from other countries should suffice to highlight how anthropological research can be a crucially important component of development programmes like dam building. Nearer home, in India, the only project of this type where I understand an anthropologist is conducting his private research is the Beirasuil Hydel Project in the Chamba Valley of Himachal Pradesh. I am told by a friend that here Professor Roger Keesing of the Australian National University is conducting a longitudinal study of the impact of the project on the surrounding populations. I have no more information about the nature of research.

Population displacement and resettlements are unavoidable if major dams happen to be sited in inhabited areas. The inhabitants will then be forced to relocate. It is also reasonable to expect that these people will experience complex problems of adjustment in the new habitat. But adequate provision to contain the likely problems can be made through planning based on research. The current proposals for water resource exploitation and management in the Northeast thus have genuine scope for incorporating anthropological research.

The harnessing of the water resources of the Brahmaputra and its tributaries, and the feasibility of constructing high dams in the north-eastern region have been considered, on and off, by various governmental agencies since the 1950s. As early as 1957, the Government of India-appointed Technical Committee on Advisability of Constructing High Dams in Seismic Zones submitted its report which considered the feasibility of storage dams on 15 major tributaries of the Brahmaputra Basin. However, it was in the 1980s that some of these proposals appeared to be taking a more definite shape. The newly constituted Brahmaputra Board has finalised and, some months ago, submitted to the Centre feasibility reports on two major dam projects across the Dibang and Subansiri Rivers, respectively. The Board is also organising investigations on a more definite footing than any time previously about the prospects of damming the Noa Dihing, the Dibang, the Lohit, etc. Besides, the Brahmaputra Board, the North-Eastern Council and the concerned States and Union Territories are pursuing relatively minor hydroelectric dam construction proposals on rivers such as the Kapili, the Doyang, the Tipaimukh, the Ranganodi and the Kameng or Jia Bhoroli. Of these, the Kapili hydroelectric project, located at Garampani in Meghalaya on the Assam–Meghalaya border, is in the final stage of completion, the work being undertaken by the NEEPCO. The two dams of this project on the Kapili River and the Umrong stream create two reservoirs. But I am told that the area is a practically virgin forest tract and there are no villages affected by the dam.

Despite my efforts, I have so far not been able to see any of the survey reports on the feasibility of dam construction across the different rivers that are now being considered in our region. However, from talks with various people as well as careful scanning of newspaper reports and articles it appears to me that the impact of dam on local inhabitants receives attention as a mere technical question in the reports. On the population front, the perspective of enquiry in the feasibility surveys appears to be quantitative and not qualitative. Thus, the reports do seem to indicate how many hectares of arable land are going to be submerged, the size of population likely to be affected and so on. But there are important social, cultural and psychological issues which a survey designed and undertaken by engineers cannot contemplate and cover. To my knowledge, none of the surveys so far completed in the Northeast have any real social science input.

Let us briefly examine a few proposals to see what the issue is from the anthropological point of view. By far the colossus among the proposals is the Dehang Project which is claimed to have a power-generating potential of a staggering 20,000 MW. The Dehang, which is one of the three main sources of the Brahmaputra in the plains (the other two being Dibang and Lohit), is known as the river Siang among the Adis of East and West Siang districts of Arunachal through whose territory it flows before entering the Assam plains. The nearly 1,000-feet (300 m)-high dam across this river is planned at a spot

29 miles upstream of Pasighat in East Siang District. It has been estimated that the dam will submerge a tract of land some 49,000 hectares in area (i.e. 121,079 acres or 189.18 sq. mile or 490 sq. km), in both East and West Siang districts. The area is occupied by Gallong, Minyong and Pdam sections of the Adi group. Long stretches of motorable roads will go under water and with these arable land and village sites of the tribesmen, who showed a tendency in recent decades to build villages by the road side. Even the township, which is some 55 km west of the dam site, will be inundated. Since the 1950s, considerable tracts of land in the valleys of the Siang, Siyum and other streams have been terraced and converted into wet price cultivation field by the tribesmen. The potential relocatees have already voiced their strong opposition to the dam fearing the loss of land and village sites. Even the Arunachal Pradesh Legislative Assembly took a position opposing the dam in its Monsoon Session of 1982. The Chairman of the Brahmaputra Board, however, came out with the assurance in his press conference of 23 June 1983 that though the two dams across the Dehang and Subansiri would submerge some 68,000 hectares of land of which 6,000 hectares are cultivated, the Board would provide 120,000 hectares of irrigated terraced land in the catchment areas to resettle the people displaced by the dam. The Board estimates that the 'incomes of the people of the area would increase by at least five times of what it is at present' (The Sentinel, 24 June 1983). The people, however, say that their developed land is dearer than gold. 'Gold can be bought but arable land in this area of steep rocky mountains, deep gorges is a God-gift and, therefore, is priceless', said an editorial in the government-controlled Arunachal News of 30 September 1982.

The issue before us is who is going to investigate the complex social, psychological and cultural aspects of the proposed relocation and when. Obviously, the sociocultural and associated ecological implications are ignored so far or, at best, considered incidental by those in authority for casting the project into shape.

The problems of relocation apart, the construction of such a major dam will involve large-scale influx of labourers, technicians, engineers, administrators, etc. into this interior hill area. The likely impact of this abrupt influx of a large number of outsiders who will perhaps stay for eight to ten years of the construction phase ought to be charted out in detail and plans made for avoiding adverse effects and social tension. The Kariba project, for example, necessitated the building of a whole new township to accommodate about 10,000 labourers, engineers and permanent operating staff and their families. The bi-national Itaipu Dam on the Parana required up to 40,000 men to work over eight years. The Dehang project will likewise involve, perhaps, thousands of workers and supporting population – always a source of potential tension in the north-eastern hills. Consider, for example, the reasons for opposition to the proposal for extension of railways to Binyhat by a section of the Khasi people (The Telegraph, 14 September 1983).

As for the other major project, namely, the Subansiri project, the dam site is at a place called Gerukamukh across the Subansiri River in the Upper Subansiri District of Arunachal Pradesh. The 853-feet (260 m)-high dam here will submerge about 19,000 hectares of land. My homework indicates that this will be in the area occupied by the Hill Miri tribesmen and perhaps also some Nishi tribesmen. But I have no information on the number of villages that are likely to be affected directly.

For the Noa Dihing, Dibang and Lohit Rivers, investigation divisions have recently been established by the Brahmaputra Board to collect information and to take up further technical investigation. My contention is that properly designed and sensitised socio-ecological investigations should form a part and parcel of the investigations for the feasibility reports themselves. Mere demographic survey of a preliminary kind, undertaken as a sort of last item in the agenda prior to the construction phase, cannot be a substitute for relocation or impact study.

As I was preparing this note, news came over the radio (evening regional news, 19 October 1983) that the NEEPCO has prepared and submitted the final feasibility report on the Ranganodi Hydel Project in Arunachal to the sponsors, namely, the NEC. The Ranganodi is the Assamese name for the river which flows through the Lower Subansiri District of Arunachal Pradesh at Panyor. I have no information yet of where the dam is sited, but know this much that the Panyor region is the territory of Nishi tribesmen. One would like to know if the report bothered itself with the socio-ecological questions and the impact of the dam on the local inhabitants. The same holds true for the Kameng Hydro-Electric Project in the West Kameng District of Arunachal Pradesh and the Doyang Project, the dam for which is sited in the Wokha District of Nagaland. A newspaper report attributed to S. R. Shah, Chairman, NEEPCO, says that all the projects undertaken in the north-eastern region are implemented taking into account the need to maintain ecological balance. But the report is silent on the possible impact on the human or sociocultural front.

In one instance, at least of water resource management, I find some governmental awareness of research by trained personnel as a part of the planning process. The instance is the Popum Poma Watershed Management Scheme of the NEC. The Planning Commission had specifically advised the council to involve university personnel in anthropological and ecological research to chalk out the likely impact of the scheme on the flora, fauna and inhabitants of the region. The Popum and the Poma are two hill rivers which join up some 20 km north-west of New Itanagar in the Lower Subansiri District of Arunachal Pradesh. This is a Nishi-inhabited area. The NEC had written to the university in this region in 1982 seeking involvement in research but nothing definite has shaped up yet.

To sum up, indigenous tribal communities with limited access to modern education and technology are particularly vulnerable to socio-cultural and psychological strains associated with abrupt alteration of habitat, resettlement and diverse outside pressures posed by large-scale influx of outsiders in their midst. Special strategies need to be devised to enable the people to cope with the immense externally induced pressures.

Irreversible problems may otherwise be generated for which no remedy will be found afterwards.

There is already a corpus of research findings from different parts of the world that have led to the formulation of theoretical propositions about how indigenous people and prospective relocatees react to sudden pressures such as dam construction. All this background is available to the anthropologists for designing research to aid sound policy formulation in our region.

Note

1 This chapter was presented at a seminar in Gauhati University, 25–26 October 1983 by Professor Bhagabati. In view of its continuing relevance even after nearly 33 years, it is being reproduced here with his kind permission.

References

Fahim, H. M. 1981. *Dams, People, and Development: The Aswan High Dam Case*. New York: Pergamon.
The Sentinel, 24 June 1983.
The Telegraph, 14 September 1983.

3 Water quality in Assam

Challenges, discontentment and conflict

Runti Choudhury, Anjana Mahanta and Chandan Mahanta

As its quantity and quality are both being increasingly challenged in a rapidly degrading planet, water, the life-sustaining resource, is perhaps predictably becoming a cause of discord and conflict with the issues of availability, distribution, security and safety – all coming under threat. In the water tower of the Himalayas, transboundary water concerns, unbridled hydropower development, rights over water, water diversion, unequal distribution, and now serious surface and groundwater contamination issues are being increasingly reported, documented and discussed in recent years, often also in the context of fuelling social unrest. Conflicts arising out of quality degradation of surface and groundwater have somehow not been discussed as visibly as the conflicts arising out of other water-related issues, although arguably it is just a ticking time bomb.

Contamination of water by arsenic, fluoride and other metals and metalloids, along with bacteriological contamination, has raised widespread concerns regarding drinking water security and safety in relatively unsuspected States like Assam, leading to resentment and mistrust particularly towards government bodies entrusted with the responsibility of water safety and security. The relative intangibility of quality dimension of water has often remained the cause of apparent inaction. This chapter presents the implicit role of water contamination in conflicts related to water security and safety in Assam.

Introduction

Contesting claim over natural resources harbouring conflict is no longer uncommon (Mbonile 2005). Conflicts over water, a critical ingredient for human survival, are becoming more obvious. Uneven distribution has compelled humans to resort to storing excess water for use during deficiency and scarcity (Armbruster *et al.* 1986; Ohlsson 1997). Unfair practices including such hoarding have the potential to trigger conflicts. The simmering, and increasingly demonstrated, conflicts arising out of hydropower development around the world are examples of unmindful overemphasis on sectoral development of resource aggravating social unrest. However, while conflicts over quantity of water are well established, water quality-related dissensions such as around arsenic and fluoride contamination of drinking water have been emerging as a major concern only recently (Sultana 2011).

Assam (89°42′ to 95°16′ E and 24°08′ to 28°09′ N), with a population of 31.2 million (Census of India 2011), covers a total geographic area of 78,438 sq. km (2.4 per cent of India), consisting of 23 districts and 219 blocks, each block further divided into gram panchayats, villages and habitations. The greater part of Assam lies within the Brahmaputra Valley, while the southernmost part lies in the Barak Valley, separated by the Central Assam range. The inselbergs (isolated hills, knobs, ridges and small mountains that appear to rise abruptly from a plain or gently sloping surrounding) located in the central districts south of the Brahmaputra are distinct features of the area. The Brahmaputra Valley is about 800 km long and 130 km wide. With 720 km of its 2,890 km length (Singh 2004) lying in Assam, the Brahmaputra River is a lifeline.

Notwithstanding a colossal monsoon surge, groundwater continues to play a greater role in meeting drinking water demands in rural Assam. Tube wells and dug wells are the most common methods to acquire water. A network of Piped Water Supply Schemes (PWSS) run by the Public Health Engineering Department (PHED) is the major water supply mechanism, supplemented by both public and private spot sources predominated by an assortment of hand pumps. A PWSS generally taps relatively deeper aquifers (~100 m) than the hand pumps (<50 m), the raw water being treated through aeration and slow sand filtration. Besides high concentration of iron (1–45 mg/L), fluoride contamination and the relatively recent discovery of elevated arsenic contamination in groundwater have greatly raised concerns regarding the health safety aspect of drinking water, leading to growing community resentment. As more and more areas are being found to be affected by some kind of contamination or the other, it is turning out to be an emerging concern as an ingredient of imminent water conflicts (Map 3.1).

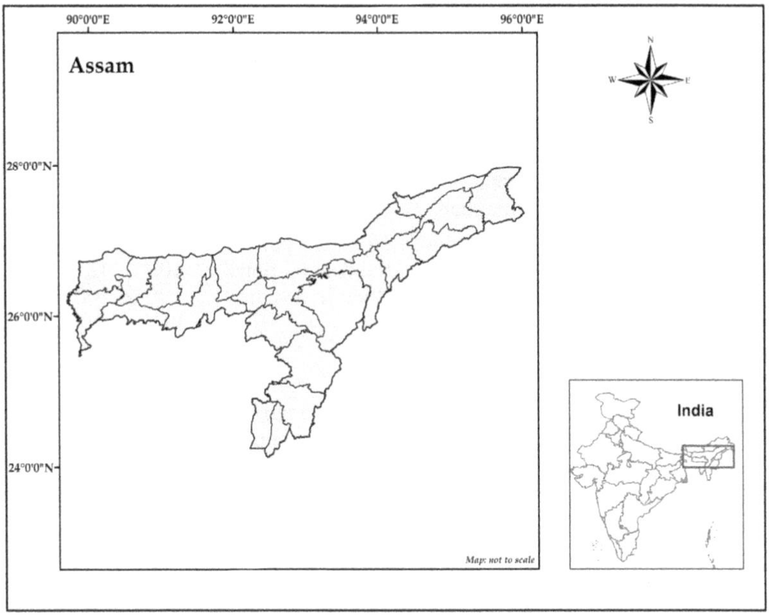

Map 3.1 Assam in Brahmaputra Valley and its districts
Source: SOPPECOM, Pune

Arsenic contamination of groundwater

Groundwater arsenic contamination in Brahmaputra floodplain is a relatively recent discovery (Table 3.1). Preliminary case of arsenic contamination of groundwater in Assam was reported by Singh (2004), with high concentration reported from Jorhat (194–657 µg/L) and Golaghat (100–200 µg/L) districts located on the southern bank of Brahmaputra, Dhemaji (100–200 µg/L) and Lakhimpur (50–550 µg/L) districts on the northern bank. Chakraborti *et al.* (2004) reported 26 per cent of 137 hand pumps sampled in two districts having arsenic concentration above the Bureau of Indian Standards (BIS) permissible limit of 50 µg/L. Based on studies conducted in Darrang and Bongaigaon districts on the northern bank of Brahmaputra, Enmark and Nordborg (2007) found concentration of arsenic ranging between 5 and 606 µg/L. Chetia *et al.* (2008), based on studies in Golaghat District on the southern bank, reported arsenic concentrations of up to 73 µg/L

Table 3.1 Concentration ranges of arsenic in different parts of Assam State in the Brahmaputra Valley

Author	*Region*		*Concentration range of arsenic in μg/L*
Singh (2004)	Nagaon Jorhat Golaghat	Southern part of the Brahmaputra River in Assam	481 to 112 194 to 657 100 to 200
	Lakhimpur Nalbari Dhubri Darrang Barpeta Dhemaji	Northern part of the Brahmaputra River in Assam	50 to 550 100 to 422 100 to 200 BDL to 200 100 to 200 100 to 200
Enmark and Nordborg (2007)	Darrang and Bongaigaon	Northern part of the Brahmaputra River in Assam	5 to 606
Chetia *et al.* (2008); Chetia *et al.* (2011)	Golaghat	Southern part of the Brahmaputra River in Assam	BDL to 73
Mahanta *et al.* (2008)	All over Assam		BDL to above 300

BDL; below detection limit

Source: Compiled by authors

in three blocks. In a comprehensive arsenic screening and surveillance programme carried out in 76 blocks of 18 districts in Assam, 29 per cent out of 56,180 public sources were found to have concentrations above the World Health Organization (WHO) guidelines of 10 µg/L, and 8 per cent sources were found to have concentrations above the BIS limit of 50 µ/L, exposing an estimated 8,47,064 population to the risk of arsenic contamination (Mahanta *et al.* 2008). Ironically, here has been limited dissemination of this knowledge to the affected public, apparently to avoid panic, and provision of alternative safe sources is likely to be a time-taking process. Since an unknown yet large number of private sources could not be covered under the screening exercise, a further layer of risk remains unexplored.

Fluoride contamination of groundwater

Fluoride contamination was reported in Assam in May 1999 in the Tekelangjun area of Karbi Anglong, where fluoride concentrations were found to be 5 to 23 mg/L (Kotoky *et al.* 2010). Subsequently,

Table 3.2 Concentration range of fluoride in different regions in Assam

Author	*Region*	*Concentration range of fluoride in mg/L*
Paul (2000)	Karbi Anglong	5 to 23
Das *et al.* (2003)	Guwahati	0.18 to 6.88
Borah *et al.* (2010)	Darrang	0.01 to 0.98

Source: Compiled by authors

more areas in Kamrup, Nagaon and other fluoride-affected districts were explored and prevalence of fluorosis in these areas was reported (Table 3.2, Map 3.2) (Kotoky *et al.* 2010; Chakravarti *et al.* 2000; Sushella 2007; Thakuria 2007, 2010). The affected people were not just exposed to a serious health threat; their socio-economic lives were affected permanently due to inability caused by crippling fluorosis rendering them completely unproductive. In order to look after crippled parents, children were forced to give up schooling, thereby losing the opportunity to make any comeback. The trauma and helplessness is something seen to be believed. While full-blown cases of arsenicosis have not been reported till date, the prevalence of fluorosis – both dental and skeletal – has been quite well-documented and reported time and again since the first discovery in 1999. Efforts have been initiated in certain affected regions to provide fluoride-free alternative sources, but a large section continues to suffer.

Conflict

Water-related conflicts can emerge at different geographical locations and on different scales. Some conflicts such as those related to hydropower can be broad based and adequately engaged with both in terms of impacts and visibility through public protests, while those arising out of water quality and quantity are localised and less visible. Much is known today about transboundary water issues, yet numerous conflicts arising out of local water issues are often little known.

An implicit conflict over the issue of safe drinking water is likely to be fuelled primarily by issues of water availability itself. Groundwater arsenic and fluoride contamination in the Brahmaputra floodplain has aggravated the living conditions of the people in the contaminated areas. Inadequate management strategies coupled with issues of improper maintenance of alternative sources and lack of smooth functioning have caused great resentment among the inhabitants in these areas. Although not as vociferous as the conflicts regarding

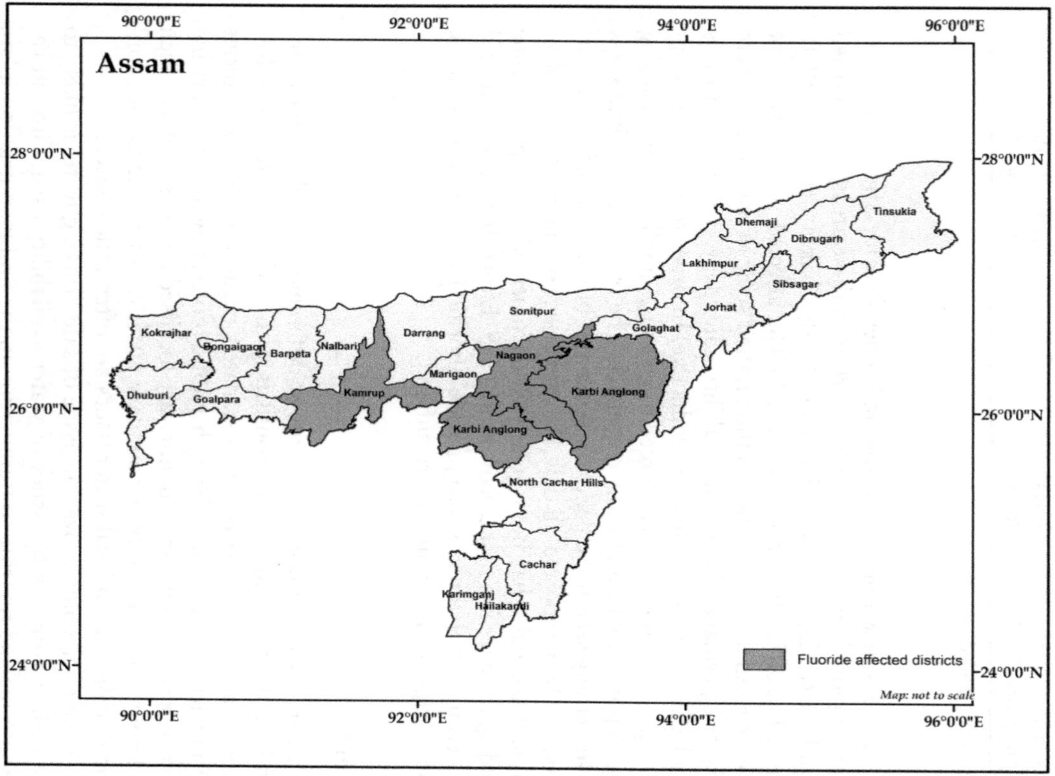

Map 3.2 Districts severely affected by groundwater fluoride contamination in Assam

hydropower dams, people's discontentment is expressed in seminars, public forums, widespread newspaper coverage, and regular reports in the electronic media. Since these issues occur on a local scale, and the exposed communities rarely have any political connections, the resulting conflicts remain rather dormant.

Water quality can cause conflict between those who cause it and those who suffer. This situation is further aggravated when the linkage between health and water quality is more conclusively established and reported. With a relatively high occurrence of cancer in the State of Assam, it can perhaps be argued, though not conclusively established, that the heavy metals carried down from the acidic waters of the rivers and streams from the upstream coal mining areas of Meghalaya could actually be responsible for the increased incidence of cancer. Once established with evidence, interstate conflict over water quality issues may perhaps become imminent. Linkages between fluoride contamination and fluorosis and that between arsenic and arsenicosis are already well established. Crippling fluorosis in the Tekelangjung areas in Karbi Anglong District, Nagaon and Kamrup districts is an evidence of fluoride manifestations, while no cases of arsenic manifestations have been reported so far. In both cases, however, conflicts between affected communities and drinking water supply agencies appear to be an obvious possibility that can lead to ugly social unrest.

A recent study by Sultana (2011) on arsenic contamination in groundwater in Bangladesh has highlighted how drinking water contamination has resulted in new meanings and realities for the access, use and conflicts in the micro-practices of water in everyday life. Arsenic contamination in Bangladesh has thus created a scenario where safe water control has become both a status symbol and a source of power (Sultana 2007). Such scenarios in Bangladesh and West Bengal have been worked out due to the fact that arsenic occurrence in these areas was discovered much earlier. Also, the fact that arsenic manifestations in these areas are well reported has made research studies from a socio-economic perspective possible.

The issue of arsenic contamination in the Brahmaputra floodplains in Assam is a relatively recent discovery. With the gradual passage of time, 29 per cent of sources which are now reported to be contaminated with arsenic above the WHO limit of 10 ppb, and 8 per cent sources which are contaminated above the BIS limit of 50 ppb, may perhaps change and newer areas of contamination may possibly be exposed. Spatial heterogeneity of arsenic occurrence and shallow aquifers being highly contaminated make our situation similar to that of Bangladesh and West Bengal, the only difference being the fact that so far no cases of arsenic manifestation have been reported in Assam. Considering the situation in Bangladesh and West Bengal, it is possible that implicit conflict is likely to emerge out of the inaccessibility of safe water, which is again a function of land ownership, ownership of tube wells, socio-spatial location in relation to a tube well, membership in a water committee or kinship and/or patron–client relations that enable access to safe water.

In the case of Assam, although the region shares some benefits in terms of being a water surplus region where there is a sense of harmony between different communities, the possibility of conflict scenarios arising out of water quality issues cannot be ruled out.

Issues

With communities facing the impacts of fluoride and arsenic contamination in the region, a growing sense of resentment has developed. Government initiatives to provide communities in the affected areas with alternative sources of drinking water are underway at times, but there is a gap between demand and supply which seems to have increased the resentment among the communities further. Credible or quantified outcome in easily understandable terms has been lacking, and the PHED has been blamed for apparently ignoring the quality aspect of water since it has adopted a supply-driven mechanism instead of demand-driven provision for drinking water. All existing groundwater sources need to be examined in case there has been not only a degradation in quality, but also severe contamination due to arsenic and fluoride.

The discontentment among the affected communities has increased because of the failure of the government machinery, inadequacy of mitigation initiatives and perhaps because the people are not the actual beneficiaries of alternative solutions. Most affected people hardly manage two square meals a day and have little capacity to alleviate the issue of fluoride and arsenic contamination at the individual level. Obviously, it is only a matter of time before the issue will bring people out on the streets.

Government initiatives for mitigation in some of the hotspot areas are underway, wherein surface water sources are being used to provide safe water to the affected areas. The recent laying of the foundation stone for a water supply project by the PHED in the greater Dokmoka Tekelangjung areas in Karbi Anglong to provide fluoride-free safe waters among the affected communities and the grandiose Titabor PWSS in the Jorhat District for providing arsenic-free safe water are

Box 3.1 Health Implication of fluoride contamination in Assam

In spite of more than a decade since the discovery of fluoride in Nagaon District in Assam, the situation still continues to worsen. Health implications due to consumption of fluoride-contaminated drinking water have impacted both social and economic lives of communities. In some areas, up to three generations of people are reported to suffer due to fluoride contamination. Although initiatives to provide alternate safe drinking water have been undertaken, the benefit of such initiatives remains questionable. These in turn led to a series of protests by locals and nongovernmental organisations (NGOs) in the affected areas.

Tapotjuri, one of the many villages under Akashi Ganga gram panchayat, has been doomed by fluoride contamination in groundwater since 1990. In a study conducted by Akashi Ganga Multipurpose Social Welfare Society, a local NGO, 925 people in the Akashi Ganga area are reported to have deformed arms and legs, body ache and dental problems due to fluoride toxicity. Such situations without stringent measures for corrective actions are reported to aggravate the situation.

Source: Mitra (2013)

some initiatives by the government. However, the irony of the situation is that there are accusations of malpractice and poor quality of workmanship in recent times that are adding newer dimensions to potential conflicts. The conflicts between the affected communities and the government occur because:

1 The real beneficiaries of the alternative drinking water sources are perhaps not the ones who are genuinely affected.
2 Public awareness about the fluoride and arsenic issues or rather the water quality issues is not adequate as of now.
3 The lack of dependable technologies for detection and treatment of the clinical manifestations (arsenicosis, fluorosis, etc.) due to consumption of arsenic and fluoride-affected groundwater complicates the health crisis further.
4 In most cases, fluorosis and related diseases often go unnoticed, leading to other ailments which are neurological, muscular or gastrointestinal in nature.

5 Arsenic and fluoride contamination bears a direct link to poverty. Therefore, the impacts of these contaminants seem to further aggravate the socio-economic lives of the affected people.
6 The government is apathetic towards debilitating fluorosis.
7 The government's mitigation programmes are not transparent.
8 Media reports at times tend to intensify the issue through provocative coverage.
9 There is a lack of proper facilities to assess whether the water is potable or not.

Inadequate mitigation

Even after 13 years since the discovery of fluoride in the groundwater of Assam in 1999, the region does not have proper healthcare centres to facilitate the treatment of patients of fluorosis. The affected people, even after being aware of their health conditions, cannot get treated due to the lack of treatment facilities. Sporadic efforts to provide alternative water sources have delivered little.

People affected by arsenic and fluoride are usually from disadvantaged communities. With their limited capacity for the prevention of arsenic and fluoride contamination, they depend largely on the government for a solution. As has been witnessed in Bangladesh, arsenic contamination being a health risk is also a risk to the economy. With an approximate total population of 26.64 million in the Northeast, and 18 out of 27 districts in Assam being affected with high levels of arsenic concentrations, the number of people exposed to the risk is large, and the crisis serious.

Conflict around water quality is only dormant

While the conflicts on water quality have manifested in the form of local protests and expressions of resentment, the implicit conflict situations around water quality, though discernible, are not fully asserted. Since such locations are scattered all over the State, and inhabitants often neither have a knowledge-based evidence backup nor sufficiently strong political voice, the conflicts do not reach a scale such that they are distinctly visible. Yet, discontentment continues to prevail, and it is the local officials who are often challenged and insulted.

One of the major steps to minimise public discontent and simmering conflicts would be to generate adequate public awareness. As 'ignorance is the mother of all evil', focussed awareness programmes could perhaps greatly minimise the gap in understanding currently prevailing concept of water quality, often attributed to only colour and taste, among the communities. Information dissemination should not remain confined to the cause and impacts of fluoride and arsenic contamination. It should also encompass aspects such as major steps undertaken by the government for mitigating water quality problems. This will ensure the transparency of mitigation programmes, and the accountability of those implementing them.

The conflicts and sense of resentment among the people are often attributed to government apathy. There is a need to reduce such indifference. Effective governance can address problems of water stress, for example, by improving storage, preservation and water quality (Gizelis and Wooden 2010). Apart from drinking water supply projects, public health centres with proper detection and treatment facilities need to be set up in each affected district. Experts from other fluoride- and arsenic-affected areas experienced in the detection and treatment of fluorosis and arsenicosis should also be involved in the training programmes.

Reports of fluorosis in Assam have already been well documented, while no such reports of arsenicosis are available. With the growing risk of arsenic in the region, the government should implement thorough health check-up campaigns for arsenicosis at least in the major arsenic-affected areas. Such initiatives in terms of the government would be appreciated and welcomed by the people. In providing alternative drinking water sources, apart from the development of PWSS, large rainwater harvesting schemes and development of indigenous water storage techniques with community involvement also have to be considered. The community has to be consulted before providing alternative solutions. The people can be entrusted with the responsibility of operation and maintenance of such alternative sources, so that they have a sense of ownership.

Eventually, a conflict needs to be resolved. For both tangible and implicit water quality conflict, a conflict resolution mechanism would call for improved coordination between the communities and the government. Local capacity must be enhanced to ensure water safety. Conserving available safe water resources followed by conjunctive use of these resources is a desirable step in water-rich regions like Brahmaputra Valley.

Poor maintenance of primary drinking water infrastructure, namely PWSS, rainwater harvesting structures, etc. in areas where the groundwater source is contaminated, often have been a major factor of failure of schemes, which in turn leads to major resentments among communities charging emotions. The issue of poor operation and maintenance can best be resolved through a mechanism involving a complete community participation from planning to implementation of local water treatment and supply infrastructure through a life-cycle approach. For instance, the emerging National Rural Drinking Water Programme (NRDWP) initiatives for communities to be entrusted with the responsibility of managing rural drinking water supply systems through village water user committees is a potential step to diffuse tensions around poor water quality supplied to stand posts and households.

References

Armbruster, B. B., C. L. Mitsakos and V. R. Rodgers. 1986. *Earth's Geography*. Lexington: Glinn Company.

Borah, K. K., B. Bhuyan and H. P. Sarma. 2010. Lead, arsenic, fluoride and arsenic contamination in the tea garden belt of Darrang District, Assam, India. *Environmental Monitoring and Assessment*, Springer, 169(1–4): 347–352.

Chakraborti, D. *et al.* 2004. Groundwater arsenic contamination and its health effects in the Ganga-Meghna Brahmaputra plain. *Journal of Environmental Monitoring*, 6(6): 74N–83N.

Chakravarti, D. *et al.* 2000. Fluorosis in Assam, India. *Current Science*, 78: 1421–1423.

Chetia, M. *et al.* 2008. Groundwater arsenic contamination in three blocks of Golaghat district of Assam. *Journal of Indian Water Works Association*, 40(2): 150–154.

Chetia, M. *et al.* 2011. Groundwater arsenic contamination in Brahmaputra river basin: A water quality assessment in Golaghat (Assam), India. *Environmental Monitoring and Assessment*, 173: 1393–1398.

Das, B. *et al.* 2003. Fluoride and other inorganic constituents in ground water of Guwahati, Assam. *Current Science*, 85: 657–661.

Enmark, G. and D. Nordborg. 2007. *Arsenic in the Groundwater of the Brahmaputra Floodplains, Assam, India: Source, distribution and release mechanisms*. www2.lwr.kth.se/Publikationer/PDF_Files/MFS_2007_131.pdf.

Gizelis, T. I. and A. E Wooden. 2010. Water resources institutions and interstate conflict. *Political Geography*, 29: 444–453.

Government of India. 2011. *Census of India 2011*. New Delhi: Office of the Registrar General & Census Commissioner. www.censusindia.gov.in.

Kotoky, P. *et al.* 2010. A fluoride zonation map of the Karbi Anglong district, Assam, India. *Research Report*, 43: 157–159.

Mahanta, C. *et al.* 2008. *Source, Distribution and Release Mechanisms of Arsenic in the Groundwater of Assam Floodplains of Northeast India*. Proceedings of the World Environmental and Water Resources Congress, Environmental and Water Resources Institute (EWRI) of the American Society of Civil Engineers, Honolulu, Hawaii, 12–16 March 2008, pp. 1–19. doi: 10.1061/40976(316)78.

Mbonile, M. J. 2005. Migration and intensification of water conflicts in the Pangani Basin Tanzania. *Habitat International*, 29: 41–67.

Mitra, N. 2013. Fluoride poisoning leaves villagers in deep water. 27 September 2013. http://timesofindia.indiatimes.com/city/guwahati.

Ohlsson, L. 1997. *Water Scarcity and Conflicts*. Department for Peace and Development, University of Goterburg, Pangani Basin Water Office. Tanga: Pangani Basin Authority Annual Report.

Paul, A. B. 2000. *High Fluoride in and Around Karbianglong District, Assam: A Case Study*. 32nd Annual Convention of Indian Water Work Association, pp. 38–41.

Singh, A. K. 2004. *Arsenic Contamination in the Groundwater of North Eastern India*. Proceedings of the National Seminar on Hydrology, Roorkee, India.

Sultana, F. 2007. Water, water everywhere but not a drop to drink: *Pani* politics (water politics) in rural Bangladesh. *International Feminist Journal of Politics*, 9(4): 1–9.

Sultana, F. 2011. Suffering for water, suffering from water: Emotional geographies of resource access, control and conflict. *Geoforum*, 42 (2): 163–172.

Sushella, A. K. 2007. *A Treatise on Fluorosis*. New Delhi: Fluorosis Research and Rural Development Foundation, pp. 15.

Thakuria, N. 2007. Struck by Fluorosis. *India Together*, 9 January 2007. www.indiatogether.org/2007/jan/hlt-fluor.htm.

Thakuria, N. 2010. Another underground disaster. Terra Green, July 2010. http://terragreen.teriin.org/index.php?option=com_terragreen&task=detail&section_id=717&category_id=8&issueid=38.

4 Harnessing energy potential in fragile landscapes

Exploration of conflicts and emerging issues around hydropower developments in Sikkim

Ghanashyam Sharma and Trilochan Pandey

Background

Sikkim Himalayas is a provider of enormous ecosystem services to the rest of the downstream which contributes to the regional economy and its human development. The glacier-fed rivers, the Teesta and the Rangit, and their system of tributaries contribute to human well-being and prosperity, providing water for its unique traditional integrated agriculture systems, hydropower generation, tourism and allied industries in the riparian Indian States of Sikkim, West Bengal and Bangladesh (Khawas 2015). The landscape is considered as the abode of deities called the *Beyul-Demazong* (the hidden fruitful valley), cultural landscape (Sherpa 2003) or the *Denzong* (the valley of rice) with an intrinsic cultural and belief system embedded deeply in the *Lepcha*, *Bhutia* and *Nepali* indigenous communities (Sharma *et al.* 2012). Natural cultural landscapes are the manifestations of the symbolic identity of the traditional indigenous communities and are the expressions of natural and ethno-cultural characteristics of multifunctional resource management. The Sikkim Himalayas has unique and diverse ecosystems from subtropical (>300 m) to trans-Himalayan (5,500 m) agro-climatic zones with rich sociocultural, socio-ecological and biodiversity values associated with it. The area is a repository of unique, globally significant biodiversity and ecosystems that provide ideal habitats for the survival of flagship wildlife and keystone species. Agro-biodiversity is equally rich in the human-managed traditional agroecosystems along the advancing altitudes (300–4,500 m) (Sharma and Dhakal 2011).

Sikkim is largely a subsistence agriculture-based economy in which almost 64 per cent of the population are still dependent on agriculture

(SAPCC 2011) and contributes 17 per cent to the gross state domestic product (GSDP). Lately, concerted efforts are being made by the government to develop eco-tourism, certified organic agriculture, pharmaceutical companies and hydropower projects to maximise State revenues. There is tremendous scope to export and sell power generated through hydropower projects over River Teesta and other tributaries and streams (Planning Commission 2008).

As per the Vision Document of the Energy and Power Department, Government of Sikkim (2010), the total annual revenue generation from the hydro project expected to be commissioned by 2015 is around Rs. 1,292 crore, which is now being revised to Rs. 1,500 crore according to the latest vision document of the same department published in 2014. In addition to this, the revenue from existing projects as well as from the proposed projects to be developed under the Sikkim Power Development Corporation (SPDC) was calculated to be around Rs. 210 crore per annum. The latest human development report (Sikkim Human Development Report Cell Government of Sikkim 2015) indicates that the hydropower projects at various stages of construction will have an installed capacity of close to 5,350 MW when completed. The State Power Policy lays down that the independent power producers will have to provide employment to the local skilled, semi-skilled and unskilled manpower as well as create local business and contract opportunities.

There are 25 projects at various stages of development across Sikkim (Table 4.1). These projects have acquired large forest and agriculture land, several sacred, spiritual and culturally important spaces along the river belts. This makes the State with the highest density of dams (4/1,000 sq. km), than any other States in the Himalayas (Pandit and Grumbine 2012).

The abrupt advent of heavy machineries for large-scale construction, mass migration of labourers from other States of the country and rapid land use changes have led to livelihood changes, internal displacement and landlessness-led migration. Such situations have created conflicts and discontent among the local and indigenous communities. This case study highlights the conflicts, community-culture-development nexus and the emerging issues raised by hydropower projects in Sikkim.

Objective and methods

This study explores the conflicts between different stakeholders and the social undercurrents around hydropower projects of East and West Sikkim. Extensive field survey and observation, focus group

Table 4.1 Hydropower projects at various stages of developments in Sikkim

S No.	*Name of the project*	*Installation capacity (MW)*	*Completed/expected to be commissioned*
1	Teesta Stage I	280	2016–17
2	Teeta Stage II	330	2015–16
3	Teesta Stage III	1,200	January 2012
4	Teesta Stage IV	520	2014–15
5	Teesta Stage VI	500	November 2013
6	Lachen HEP	210	2016–17
7	Panan HEP	280	September 2014
8	Rongnichu HEP	96	July 2014
9	Sada Mangder HEP	71	March 2014
10	Chuzachen HEP	99	February 2011
11	Bhasmey HEP	51	March 2013
12	Rangit II HEP	66	June 2014
13	Rangit IV HEP	120	June 2014
14	Dikchu HEP	96	December 2014
15	Jorethang Loop HEP	96	December 2012
16	Lachung HEP	99	October 2016
17	Bimkyong HEP	99	April 2016
18	Bop HEP	99	October 2015
19	Ting Ting HEP	99	December 2013
20	Teesta Stage IV HEP	510	Completed and commissioned
21	Tashiding HEP	97	September 2013
22	Lethang HEP	96	October 2014
23	Suntaleytar HEP	30	2014–15
24	Kalez Khola II HEP	60	2014–15
25	Rangit III	60	Completed and commissioned

Source: EPDGOS (2015)

discussions, interviews with key informants and various government and nongovernmental institutions, and rapid interactions with indigenous and local communities, official of the hydropower developers, Sikkim Government officials of the Power and Energy Department, panchayats (local self-government in India), school teachers and other stakeholders in and around the project area were conducted. Further, we recognised the value of local perception as a key source of information on changing situation due to the increasing hydropower project activities in their localities, and as such, our respondents were the primary informants to document the indicators of change due to project activities and increasing migration of labourers from the plains. Extensive field observations were made in the project areas to understand general ecology and environment. Further, we reviewed the available literature, policy documents of the governments and followed local media reports. We also carried out a detailed study on 'Benefit Sharing Mechanism in Storage Projects: Lessons for Region (based on National Hydropower Corporation (NHPC) Teesta Stage V Sikkim)'. This study identified the benefit-sharing mechanism developed for providing incentives to the project-affected communities along the Teesta Stage V NHPC project areas.

Conflicts and emerging issues

Hydropower generation has made important contributions to human development, and the benefits that can be derived from them are significant. The Government of Sikkim aimed to develop its economic plans to become self-reliant with adequate revenue generation by capitalising on its perennial rivers as renewable resources. Being a special category State with international border, the State effort to exploit water resources is to reduce the fiscal dependency on the Central government. The Sikkim Government's 209-paged white paper prepared by Entecsol International, tabled in the State Assembly on 30 July 2009, mentions that the State has a potential to generate 8,000 MW hydropower (Map 4.1). Sikkim has allotted around 25 hydropower projects for construction with a calculated installed capacity of 5,284 MW mainly capitalising the water of River Teesta and Rangit and their tributaries.[1] Surprisingly, as many as six power projects have been envisioned in the River Teesta supported by multiple high level dams built within 175 km distance with an estimated generation capacity of 3,635 MW (Lepcha 2013), while Teesta Low Dam (III and IV) in Darjeeling District of West Bengal before the Coronation Bridge were commissioned in December 2012.

As per Saumitra Ghosh, an environmental activist,[2] although the projects had promised for adequate compensation and jobs to the communities living in the scattered forest villages inside the gorge and in the higher valleys, mired in controversies since its operation. The 400 families were left with no compensation and are ignored; now await the submergence of their land property, homes and wayside shops that provide them livelihood. The River Teesta is now blocked in several locations and does not even have its minimum ecological flow, while the project authorities have completely disregarded geological impact that would anytime bring about a disaster by re-activating landslides all along the NH 31A (new numbering NH 10) lying above approximately 12 km long reservoir.

The most challenging socio-economic issues in Sikkim are the displacement of local and indigenous people, mostly farmers, owing to acquisition of their land by the projects through the government. A large number of farmers have become landless, and they even could not properly utilise the money they received as compensation due to lack of financial literacy.

Besides, the local population are not given much preference for employment by the power developers, even if this was guaranteed to the locals in the project plans and policies (www.sikkimpower.org). The resettlement and rehabilitation (R&R) plan envisages several incentives for the displaced families which are yet to be fully implemented and monitored.

Listing the recurrent impacts

The socio-economic and socio-ecological impacts of hydropower projects include migration, landslides, damage to houses and cultivable land, changes in household size and structure; changes in employment and alteration of income-generating activities and opportunities; alteration of access and use of land and water resources; changes in social networks and community integrity; and often, a disruption of the psycho-social well-being of displaced individuals (WCD 2000). Several cases of disappearance of springs, which are the immediate sources of drinking water, have been reported in the project sites.

Our focus group discussions and extensive field survey revealed that the strong blasting in the tunnelling process creates cracks on rocks, which resulted in disappearance of springs and groundwater. Besides, several houses have developed cracks on the walls and some have even tilted and become inhabitable in the project sites.

Sikkim is identified as geologically fragile and is part of seismic Hazard Zone IV prone to high magnitude earthquakes (Nath *et al.* 2008). The 18 September 2011 earthquake (6.9 Richter scale) caused more than 200 landslides, swept away several villages mostly in Dzongu in North Sikkim and killed more than hundred lives. The landslides and casualties as observed and studied were comparatively *more* in the hydropower project sites across the State than in non-project sites.

The questionnaire-based survey in project areas revealed various impacts such as soil erosion, landslides, incidence of floods, emergence of new diseases (dengue fever, HIV AIDS, Malaria, etc.), disappearance of springs, pollution/traffic have significantly and rapidly increased in the last ten years thus bringing conflicts, discontent, dissatisfaction,

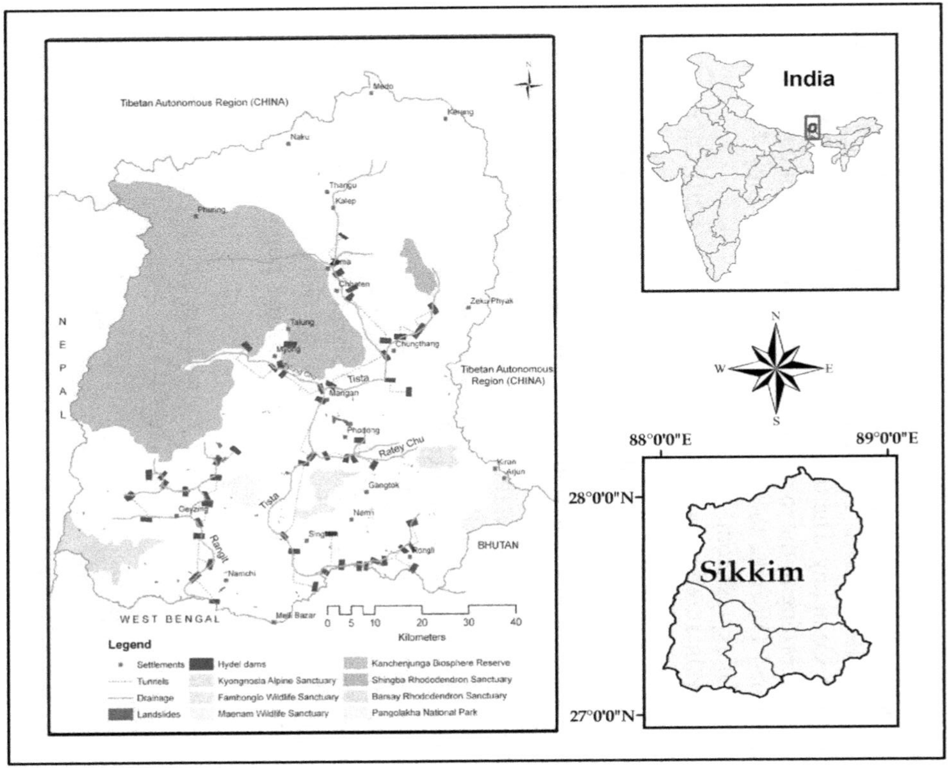

Map 4.1 Protected areas, hydropower projects, tunnelling and landslides in Sikkim

Source: The Mountain Institute, India

resentment and opposition among the local and indigenous communities. The indigenous communities reveal that they have lost their symbolic identity and ownership over the resources, environment and the aesthetic/spiritual value of their sacred landscape.

There are more cases of accidents reported due to increasing traffic and in some cases the existing roads and bridge infrastructure does not have the capacity to bear the growing intensity and load. Ten local people were killed on 19 December 2011, when a concrete steel bridge collapsed over Rangchang River in East Sikkim due to the heavy weight of a 48 wheelers truck carrying a transformer of 1,200 MW Teesta Urja Company. The truck was reported to plunge 70 feet into the river according to the local media and the communities. As per the Border Road Officials, the bridge had a capacity of only 70 tonnes (www.indianexpress.com).

Similarly, in another shocking incident of 19 June 2013, a massive hole was surfaced at Sumin Linzey village due to tunnelling work carried out by of Madhya Bharati Company (Figure 4.1). Local people reported this as the result of the use of heavy explosives for tunnelling work. This incident brought the entire village and the adjoining areas under grave risk.

Our case study in the NHPC Teesta Stage V project area shows that the traffic congestion, pollution, accidents and deaths have increased dramatically over the years once the hydropower developers initiated their massive project activities along the Teesta River. Landslides and road blockages are frequent at several locations along Rongpo-Singtam-Dickchu-Mangan-Chungthang road. The local communities living in this stretch feel that they are the worst sufferers; as such concerns are not noted and discussed in the popular discourse of

Figure 4.1 A hole on the surface due to unscientific tunnelling work

Source: Photo: http://voiceofsikkim.com/2013

impacts, rehabilitation and relief initiatives of both state and non-state actors.

Of the total hydropower projects envisioned and proposed in Sikkim, eight projects (Teesta Stage III, IV and V projects located in Panam, Rangyong, Rukel, Ringpi, Lingzya) are directly or indirectly associated with the Dzongu area while the calculated capacity of energy generation was around 2,500 MW (Subrata 2013). There was a large-scale protest of the Lepchas against the hydropower projects. Further, the Lepchas feel that the gas used in blasting has adversely affected the productivity of the cardamom by about 50 per cent. Dust pollution is rampant affecting the flowering and productivity of fruits. The Lepchas residing in the area are aware of the threat associated with the development of hydropower projects and at present are concerned for their future and their rights (Subrata 2013).

The level of negligence by the hydropower companies in Sikkim in many occasion have escaped the local political discussion (www.voiceofsikkim.com; accessed on 10 September 2013). The hydropower developers have completely violated the Environment Protection Act, Forest Acts and Policies, Biodiversity Acts, etc. of India (Lepcha 2013; Huber and Joshi 2013; Sharma and Pandey 2013). They have disregarded the culture, and rights and concessions of the local and indigenous people neglecting fragility of the mountain landscapes.

Teesta Stage V Power Station

The 510 MW Teesta Stage V Power Station, one of the six hydropower schemes in a cascade identified on the Teesta, is a run-of-the river scheme located in the East District of Sikkim. Constructed at a cost of Rs. 2,619 crore, it is one of the largest hydel power projects commissioned in Northeast India in 2008. Teesta Stage V Power Station is envisioned to work as a peaking power station, helping in stabilising the eastern grid. The annual power generation from the power station is estimated to be 2,573 Gigawatt hour (GWh).[3] The beneficiaries from the Teesta Stage V are Damodar Valley Corporation, West Bengal, Bihar, Sikkim, Orissa and Jharkhand.

The NHPC acquired 174.09 ha for construction purpose from private holdings, fully affecting 60 households (both houses and lands) and partially affecting 199 households (only land) as per the NHPC R&R plan according to the details provided to the Sub-Divisional Magistrate (SDM), East District. It submerged 48.9 ha forest land and 18.9 ha non-forest land. The website of the NHPC states that they afforested 250 ha and planted 0.8 million tree saplings.[4] The headrace tunnel had a leakage of large volume of water freely flowing elsewhere along the road. The leakage was attributed to the cracks and damage caused to the large number of aquifers above the tunnel due to high intensity dynamite disturbing the sub-surface hydrogeology of the entire area.

Fear of flooding and damage

The residents of Singtam, Rangpo and villages along the Teesta River live with fear of floods every year, mainly during monsoon. Due to intense rains, the dams release water more often and in some seasons it rises 5 feet higher than usual, flooding several houses in Adarsh Goan and Singtam (Figure 4.2). Local people have been strongly arguing that there is serious lack of information about the floods and release of water from the Dikchu dam. They also reiterated that project developers have not constructed protecting walls along the river belts especially around the human habitations.

The NHPC Stage V and the Lanco Power Project (currently a sick project) are operational in the area and caused landslides in the area affecting the National Highway 31A (now NH10) which links the rest of the State with north and south district. The project completely

Figure 4.2a Teesta River during monsoon creating threats to Singtam and Adarsh (Manpari) Busty

Source: Photo: Ghanashyam Sharma

Figure 4.2b Teesta River during monsoon creating threats to Singtam and Adarsh (Manpari) Busty

Source: Photo: Ghanashyam Sharma

destroyed the alternative road connecting the important town centre of Singtam, disconnecting the villages of Tokal Bermek and Timi Tarku. Daily public transport systems are regularly disturbed, making the villagers unable to access their nearest town Singtam for procuring consumer goods, healthcare services and to sell their agricultural produce. Several landslides are visible on both sides of the Teesta in and around the Lanco Project area. The civil works of the Lanco Teesta Hydropower Pvt Ltd went stalled for many months, due to financial burdens although a huge investment and construction work was completed. There is very less public information on this project progress but key informants have revealed that the detailed project report did not match with the ongoing work plans. Most of the regular employees of the company, mostly from the plains have left, many without salaries. The locals employed and sub-contractors are unpaid from March 2013. Such issues led to aggrieved local employees flaming the company properties in Sirwani project site on 6 April 2013 (Voice of Sikkim 2013).

None of the power developers have well-developed public information and warning system around the project areas and downstream that would arise from frequent water release from the dams for physical safety and preparedness. The general ecology along the river banks and the natural flow of river water has been completely altered over the years causing irreparable perturbations to the natural landscapes (Lepcha 2013; Huber and Joshi 2013; Sharma and Pandey 2013) (Figure 4.3).

High insecurity in Dipu Dara, near Singtam

Around a kilometre away from the NHPC Stage V powerhouse and the project office, lies a small settlement at Dipu-Dara with around 20 households. This place is residential and partly commercial, with shops and small eateries serving the travellers, and a private school Anshu Academy with more than 150 students. This settlement lies exactly above the underground powerhouse and just below the surge shaft and extended tunnel. On 5 December 2009, heavy seepage occurred from the tunnel, after the development of a crack just above the settlement. The seeped water accumulated in the roadside forest check post area which was later diverted towards the downstream *nallah*. The worried local inhabitants informed the relevant authorities at the NHPC and the State government, leading to a joint inspection in the presence of the locals.

The seepage continued for more than three months, even in the midst of grouting works in the area. Five months later, on 15 April

Figure 4.3 The Vanishing Teesta as the water flows only in the tunnels; a JCB quarrying on the river is seen in the picture.

Source: Photo: Ghanashyam Sharma

2010, heavy seepage occurred triggering a small landslide just below the old crack. The seepage was later controlled by decreasing the water flow in the tunnel by the NHPC authorities. The local people approached the authorities at the NHPC as well as the respective government departments with their problem. However, the local inhabitants feel that the situation arising from a cracked surge shaft may lead to irreparable loss of lives and property in their locality and all along the Teesta downstream.

Many houses above the 23 km road stretch have developed cracks and roads frequently get blocked due to constant landslides. People believe that the number of landslides have gone up, after the advent of the projects in the area. According to them, a number of springs have dried up as a result due to tunnelling, which leads to piercing of the aquifers within, unable to hold and release water timely and seasonally. The drying of springs has caused water shortage for drinking as well as for agriculture in the villages, below which tunnels have been constructed. The local part-time fishermen have observed that the percentage of fishes in the Teesta has gone down drastically. They cited that the present catch is only 10 per cent of the past.

Communities have reported growing danger in performing various last rites and rituals in the river Teesta. Often people have to run away leaving behind the dead body from the river banks, to get flushed by the swelling river after water is released from the dam without information. There is a serious lack of information on flow and alarm systems extending to the riverbanks where people walk to villages, and to towns like Singtam and Rangpo.

The unsettled issues of NHPC V

Residents facing damage to their homes due to tunnelling and subsequent increased seismic vulnerability of the region have had a difficult time gaining compensation for the destruction, as 'only those whose homes or lands were going to be submerged were listed' as project-affected people (Menon and Vagholikar 2004). The R&R plan enabled employment of 57 individuals out of the 60 affected households and partially affected households. However, there was growing discontent with the corporation's human resource management, as the locals who were employed at the lowest grade earlier were stuck in the same position for more than four and half years with minimal scope for career advancement irrespective of their educational qualifications. Those with low educational qualifications were given training for the higher grade, but are kept in the low grade continuously. A series of requests to the NHPC management, from the head of the project to the Chief Managing Director (CMD) level, to reconsider an upgrade, did not yield any positive response.[5]

Striking employees for proper salaries

Consequently, the agitated employees went on a strike on 31 August 2010, demanding up-gradation as per their educational qualification as agreed in the R&R plan of the project, crippling the administrative work of the project plant. The strike was also joined in and supported by the All Sikkim Democratic Labour Front (ASDLF), the frontal labour organisation of the ruling party in the State and other trade unions.

The State administration negotiated a meeting for discussing the issues concerned, which was chaired by the District Collector on 3 September 2010. This was attended by the NHPC management, represented by the Executive Director, agitating employees, the Superintendent of Police (East District), the Managing Director of the SPDC, the Assistant Superintendent of Police (East) and the Sub Divisional Magistrate. They could not come to any substantial grounds to resolve the issue. The NHPC as a committed stakeholder ensured various steps for career progression and skill up-gradation, with the contention that the decision for employees' up-gradation in the Teesta Stage V will affect other W-O employees across the country, inviting the wrath of the trade unions of the NHPC. The State representatives supporting the employees objected to this contention, observing the gross failure of the NHPC responsibility towards its employees.

The NHPC bought more time to delve into the matter at the corporate level and resolve the issue in three months. Many such cases are on the rise in the State of Sikkim owing to the initiation of several power projects. In 2011, the NHPC finally calmed down to hear the voices of its employees. The employees were absorbed in the regular positions as per their qualifications and expertise.

Power projects in the sacred *Demajong* below Mt. Kangchendzonga

The indigenous communities irrespective of the ethnic communities are the nature worshippers, the trees, plants, mountains, streams, rivers and rocks are considered sacred by them (Lepcha 2013; Sharma and Pandey 2013). The Rathong Chu, the swiftest tributary of the Rangit in the lap of Mount Kangchendzonga, originates at an elevation of 4,900 m from the Rathong glacier, carrying a rich age old tradition and culture besides providing tangible and intangible services in the catchment area. The '*Demajong* landscape' is based on the Tibetan Buddhist philosophy with clearly defined norms, and sacredness that bestows intangible benefits to the indigenous communities living in the Eastern Himalayan region (Ramakrishnan 2003, 2008) (Figure 4.4). The Hindus consider the entire landscape as *Indrakil* (the Kingdom of the Lord of Rain, the heaven) and Lepchas, the aboriginal and the primitive tribes call it *Mayal-Lyang*, the centre of the origin of the Lepchas and the treasure house of everything (Sharma and Pandey 2013). The Rathong Chu is believed to be a sacred river, full of religious rituals and practices, with its source in the nine holy lakes in the higher elevation. This religious and cultural landscape extends from the Mount Kangchendzonga peak to the subtropical agro-climatic zones within which are the natural and cultural heritage sites, historical monuments, spiritual spaces and the human-managed ecosystem diversity.

One of the interesting and holiest of all festivals is '*Bum Chu*' organised annually at the Tashiding monastery, where water collected at the confluence of the Rathong Chu and the Ringnya Chu is served to the

Figure 4.4 Puja offering at Thangu Monastery (4,300 m) to *Demajong* cultural landscape

Source: Photo: Ghanashyam Sharma

deity (Dokhampa 2003). This festival is widely attended by people from different parts of Sikkim, Darjeeling, Nepal and Bhutan.

The Rathong Chu is also not spared, even though it has an eventful history of power generation efforts and social undercurrents. Different interest groups are currently opposing the proposed three hydroelectric projects on the sacred river on the grounds of religion and culture, besides other common threats it can pose to the cultural landscape, tourism and economy. Yuksom is a base camp for Kangchendzonga Biosphere Reserve (KBR), one of the protected areas and a famous ecotourism destination. The KBR connects the Great Lhonak, Lashar Valley, and the Tsho Lahmu plateau that house the habitat of the last surviving species of mammals such as snow leopard, Kiang, Tibetan Gazelle, Himalayan Tahr, Blue Sheep, etc.

The proposed projects in the Rathong Chu are the 97 MW Tashiding HEP, the 96 MW Lethang HEP promoted by Noida-based Kalpan Hydro Pvt Ltd and the 99 MW Ting Ting HEP promoted by SMEC (India) Pvt Ltd, New Delhi.

Many organisations such as the Sikkimese Bhutia Lepcha Apex Committee (SIBLAC) and the National Sikkimese Bhutia Organisation (NASBO), and various monks associations from different monasteries, resisted the 96 MW Lethang hydropower project in Yuksom in the West District of the State and petitioned the ministry. They contested that the project was adjacent to the Kangchendzonga National Park (KNP), and the river is considered sacred to the locals, thus violating the Places of Worship (Special Provisions) Act of 1991 extended to Sikkim in 1998.

This was not the first time that such a project on the Rathong Chu was opposed and obstructed. The State government commissioned a one-man committee under Prof. P. S. Ramakrishnan, Senior Scholar from Jawaharlal Nehru University and an eminent ecologist, to carry out the fact-finding mission under the pressure from the local public and various organisations. His report raised serious concerns on such hydel projects in the 'cultural landscape' and suggested alternatives stressing on sustainable agro-forestry measures for local area development and conservation of the *Demajong* landscape (Ramakrishnan 2003, 2008). Prof. Ramakrishnan has described the region as appropriate to be declared as a 'national heritage site' where ecological considerations cannot be separated from historical, social, cultural and religious dimensions of the problem. He finds the people living in the sacred landscape are truly integrated within the landscape unit itself, in a socio-economic sense (Ramakrishnan 1998, also reiterated in 2011 in a personal communication).

Healing the wound – scraping the hydel project

As a consequence of the one man commission report, the State government had to scrap the 30 MW hydel project (Rs. 20 crore was already spent for construction) in 1997 in the Yuksom Valley (Figure 4.5). It led to employment loss of nearly 100 local individuals, employed in different capacities.

Similarly, various organisations including SIBLAC, NASBO, Concerned Lepchas of Sikkim (CLOS), Affected Citizens of Teesta (ACT), All Sikkim Educated Self-Employed and Unemployed Association (ASESUA), SAVE Sikkim, People's Forum on Earthquake and Nagarik Sangathan Sikkim (NASS) have come together to constitute Joint Action Committee (JAC) to oppose projects in Rathong Chu. Over a period of time, they have submitted memorandum to the State government to scrap the three projects – the Ting Ting (99 MW), the Tashiding (96 MW) and the Lethang (96 MW) – on the Rathong River on sociocultural and socio-ecological reasons. This was unequivocally supported by the Buddhist monks of all the monasteries of Sikkim.

Figure 4.5 The abandoned residential colony now used as block administrative centre. Many of the quarters are in a deteriorating shape.

Source: Photo: Trilochan Pandey

Such civil society organisations have questioned the veracity and authenticity of the recommendation of the high-power state committee for the Rathong Chu project. Many locals and members of minority commission believe that the committee did not address the religious-spiritual aspects associated with the anti-Buddhist Projects vis-à-vis *Denzong Neyig*.

At least 10–12 power projects have been opposed on religious-cultural grounds. Many people are worried that tunnelling activities will also damage the *Pemayangste* Monastery. The *Pemayangste* symbolises the heart plexus of the body, *Pema* means lotus, and *Yangste* means the centre, and is a part of the *Beyul-Demajong*.

Legality and closure of hydropower projects on public interest

The State government has been questioned on the legality, transparency, accountability and services of the hydropower projects and the awardees of the projects by the public agencies and also by the comptroller and auditor general (CAG) of Sikkim from time to time.

In early 2012, the Government of Sikkim scrapped six existing power projects in the larger interest of the sentiments of the indigenous communities. All these projects were at different construction phases and had occupied considerable amount of agriculture and forest area.

On 8 February 2012, the Home Department of the Government of Sikkim brought out a notification (no. 12/ Home/2012) to order the closure of hydropower projects the 99 MW Ting Ting HEP and the 96 MW Lethang HEP.

The *Dzumsa* (a traditional institution and a recognised local self-government) at Lachen and Lachung does not favour such developments, and has opposed such projects looking at the eco-cultural relation and function of balanced environments. The *Dzumsa's* as a traditional institution stake in the State policy yielded influence on the government together with support from people led to the scrapping of the 99 MW Bop, the 99 MW Bhimkyong and the 99 MW Lachung hydropower project all operational in the Lachung Chu. The Government also scrapped the 280 MW Teesta Stage I project of Lachen Chu (Government of Sikkim 2012; Secretary, Energy and Power Department, personal meeting on 28 May 2012).

The Centre for Interdisciplinary Studies of Mountains and Hill Environment (CISMHE) in its 'Carrying Capacity Studies in the Teesta Basin in Sikkim' in 2006 (www.sikenvis.nic.in) had categorically delivered the scientific evidence that such hydel projects at various locations in Sikkim that would be adverse to the environment and ecology of the region. In the biodiversity conservation front, Sikkim, the most species-rich region of India (CISMHE 2007) would have the highest dam density in the world once all the proposed hydropower projects will be commissioned, while there was a significant positive correlation between number of dams and species richness of angiosperms, birds, fishes and butterflies and as a consequence species extinction will be high in the near future (Pandit and Grumbine 2012).

The Buddhist monks and the SIBLAC had challenged the legality of the Rathong Chu HEP in the High Court of Sikkim. The apex court of the State have issued notices on environment and religious considerations to the Ministry of Environment and Forests, Government of India, National Wildlife Board, five public financial institutions, Sikkim Government and the private power developers (Sikkim Express 29 May 2012).

The Chief Minister of Sikkim has occasionally raised various questions to the 'deaf and dumb' attitude of the power developers, especially in the implementation of the R&R plan, and provision as per the agreement signed by the Sikkim Government.

Conclusion

The numerous debates on operation and impacts of hydropower developments across the State and beyond have over the time questioned the 'greenness' of hydropower systems. At a point of time when a natural seismic occurrence has shaken the stability of an already unpredictable Himalayan ecosystem, it is important to revisit the concept of hydropower projects as a green, clean and safe option (Kohli 2011). The power projects have already earned the sobriquest of *durey musa* (mountain beaver), the local people say *Darjeeling le Gorkhaland hosh is na-hosh, tara Sikkim chain todkaland bhayo* (whether Darjeeling gets Gorkhaland or not, but Sikkim has become a hole-land, i.e., the land of tunnels).

However, hydro projects can bring considerable income to the State and its people, if they are 'tried and tested' at one location within a given physical, social, ecological and economic, environment, and replicated elsewhere in a specific enabling environment. The State should develop long-term strategies and policies, taking into consideration a multidimensional approach to development. The public sector undertaking NHPC has completed its projects in due timeline, whereas other private power developers have not been able to complete their projects accordingly. Few developers such as Gati Infrastructures Ltd (Bhasmey

HEP 51 MW), Lanco energy Pvt Ltd (Teesta Stage VI HEP 500 MW), Teesta Urja Pvt Ltd (Teesta HEP III 1,200 MW) and Madhya Bharati Power Corporation (96 MW Rongnichu HEP) could not complete their projects after huge investments and acquiring large areas. As per the vision document of the Government of Sikkim, these companies were expected to complete and commission the projects before 2013 (EPDGOS 2015).

The speedy hydel project development is still on its way in the absence of effective water resource policy and governance framework. The State should build its capacity and capability to effectively manage about two dozens of power developers it has invited. The communities demand adequate compensation of their land, rehabilitation and resettlement to a safer environment, employment, health and education facilities in the affected areas. Vulnerability, deprivation, poverty, displacement and ecological crisis are the most difficult challenges in the 'Greenest State' of India. Considering the spirituality and the sacredness of the Sikkim Himalayan Landscape, the State government could declare some streams and rivers as the 'cultural heritage and sanctuary' which can serve as examples of un-perturbed natural heritage to the future generation.

Notes

1 Ministry of DONER www.mdoner.gov.in/content/hydro-power-projects#Mar12.
2 http://social-praxis.blogspot.in/2013/07/the-case-of-teesta-low-dam-projects.html, 10 June 2013.
3 Extracts from www.nhpcindia.com/Projects/English/Scripts/Prj_Introduction.aspx?vid=17 (accessed on 27 November 2010).
4 Based on Communication no. NH/TSV/HR/6075/5801/3221 to SDM (East) dated 25 November 2010.
5 Based on interaction with the employees and reference to papers exchanged between the employees and the NHPC management in 2012.

References

CISMHE (Centre for Interdisciplinary Studies of Mountain & Hill Environment). 2007. *Carrying Capacity Study of Teesta Basin in Sikkim*. New Delhi and Faridabad, India: Ministry of Environment & Forests, Government of India and NHPC.

Dokhampa, R. N. 2003. Origin of the Bumchu (Bum Chu) of Drakar Tashiding (Brag Dkar Shis Sdings). *Bulletin of Tibetology*, 39: 25–30.

EPDGOS. 2015. *Energy and Power Sector-Vision 2015*. Government of Sikkim, Sikkim, India: Energy and Power Department, p. 21.

Government of Sikkim. 2012. *Government Notification for Closure of Ting Ting and Lethang Power Projects*. No. 12/Home/2012. Gangtok: Home Department, Government of Sikkim.

Huber, A. and D. Joshi. 2013. 'Hydropower in Sikkim: Coercion and Emergent Socio-environmental Justice'. In: Das, Partha J., Chandan Mahanta, K. J. Joy, Suhas Paranjape and Shruti Vispute (eds.). *Water Conflicts in Northeast India: A Compendium of Case Studies*. Pune: Forum for Policy Dialogue on Water Conflicts in India, pp. 102–110.

Khawas, V. 2015. Dynamics of hydropower development and regional environmental security in the Teesta Basin. *Sikkim Express*. 7 June, 2015. www.sikkimexpress.com.

Kohli, K. 2011. Inducing vulnerabilities in a fragile landscape. *Economic and Political Weekly*, 46(51): 19–22.

Lepcha, T. 2013. 'Hydropower Projects on the Teesta River: Movement against Mega Dams in Sikkim'. In: Das, Partha J., Chandan Mahanta, K. J. Joy, Suhas Paranjape and Shruti Vispute (eds.). *Water Conflicts in Northeast India: A Compendium of Case Studies*. Pune: Forum for Policy Dialogue on Water Conflicts in India, pp. 73–87.

Menon, M. and N. Vagholikar. 2004. *Environmental and Social Impacts of Teesta V Hydroelectric Project, Sikkim: An Investigation Report*. Pune: Kalpavriksh Environmental Action Group.

The Mountain Institute (TMI) India. 2010. *Non-Invasive Monitoring to Support Local Stewardship of Snow Leopards and Their Prey*. Project Report Submitted to Critical Ecosystem Partnership Fund (CEPF) and Ashoka Trust for Research in Ecology and Environment (ATREE). Gangtok: The Mountain Institute India.

Nath, S. K., K.K.S Thingbaijam and A. Raj. 2008. Earthquake hazard in Northeast India – A seismic microzonation approach with typical case studies from Sikkim Himalaya and Guwahati City. *Journal of Earth System Science*, S2: 809–831.

Pandit, M. K. and R. E. Grumbine. 2012. Potential effects of ongoing and proposed hydropower development on terrestrial biological diversity in the Indian Himalaya. *Conservation Biology*, 26(6): 1061–1071. doi: 10.1111/j.1523-1739.2012.01918.x.

Planning Commission. 2008. *Sikkim Development Report*. New Delhi: Planning Commission, Government of India, and Academic Foundation.

Ramakrishnan, P. S. 1998. Ecology, economics and ethics: Some key issues relevant to natural resource management in developing countries. *International Journal of Social Economics*, 25(2/3/4): 207–225.

Ramakrishnan, P. S. 2003. *Biodiversity Conservation: Lessons from the Buddhist Demajong Landscape in Sikkim India: The Importance of Sacred Natural Sites for Biodiversity Conservation*. Proceedings of the International Workshop in Kunming and Xishuangbanna Biosphere Reserve. People's Republic of China. 17–20 February 2003.

Ramakrishnan, P. S. 2008. *The Cultural Cradle of Biodiversity*. New Delhi: National Book Trust.

SAPCC (State Action Plan for Climate Change). 2011. *Sikkim Action Plan on Climate Change (2012–2030)*. Draft Report. March 2011. Gangtok: Government of Sikkim.

Sharma, G. and T. Dhakal. 2011. 'Opportunities and Challenges of the Globally Important Traditional Agriculture Heritage Systems of the Sikkim Himalaya'. In: Arawatia Sharma, A. G., D. P. Sharma and D. R. Dahal (eds.). 2012. *Adaptive Approaches for Reviving the Dying Springs in Sikkim*. Gangtok: The Mountain Institute India, pp. 379–402.

Sharma, G. and T. Pandey. 2013. 'Water Resource Based Deployments in Sikkim: Exploration of Conflicts in the East and West Districts'. In: Das, Partha J., Chandan Mahanta, K. J. Joy, Suhas Paranjape and Shruti Vispute (eds.). *Water Conflicts in Northeast India: A Compendium of Case Studies*. Pune: Forum for Policy Dialogue on Water Conflicts in India, pp. 88–101.

Sharma, G., L. K. Rai and B. K. Acharya. 2012. 'The Socio-Cultural and Socio-Ecological Dimensions of *Demazong* Sacred Himalayan Landscape in the Eastern Himalayas'. In: Ramakrishnan, P. S., K. G. Saxena, K. S. Rao and G. Sharma (eds.). *Cultural Landscapes: The Basis for Linking Biodiversity with the Sustainable Development*. New Delhi: UNESCO and National Institute of Ecology, pp 151–168

Sherpa, L. N. 2003. *Sacred Beyuls and Biological Diversity Conservation in the Himalayas*. The Importance of Sacred Natural Sites for Biodiversity Conservation. Proceeding of the International Workshop in Kunming and Xishuangbanna Biosphere Reserve. People's Republic of China. 17–20 February.

Sikkim Human Development Report Cell Government of Sikkim. 2015. *Sikkim Human Development Report 2014: Expanding Opportunities, Promoting Sustainability*. New Delhi: Routledge.

Subrata, P. 2013. Hydro power development and the Lepchas: A case study of the Dzongu in Sikkim, India. *International Research Journal of Social Sciences*, 2(8): 19–24. www.isca.in/IJSS/Archive/v2/i8/4.ISCA-IRJSS-2013-111.pdf

World Commission on Dams (WCD). 2000. *Dams and Development: A New Framework for Decision-Making*. London: Earthscan.

5 Hydropower conflicts in Sikkim

Recognising the power of citizen initiatives for socio-environmental justice[1]

Amelie Huber and Deepa Joshi

Only a decade ago, it was argued that the old order of 'development through mega-dams . . . as shining icons of prosperity and modernity' was fading (McCully 2001: xvi). The rate of dam-building worldwide had halved, compared to its peak in the 1970s, partly due to questionable financing, and a scarcity of suitable locations for mega-water infrastructure, but also due to successful civil society advocacy on the controversial effects of dams (Khagram 2004; WCD 2000).

Today there is a resurgence of dam-building especially in emerging economies. Large dams are back on the development agenda of many national and regional governments, mostly for hydropower generation and/or water transfer to meet industrial and urban needs. The sharp reversal in the positioning of dams is particularly intriguing. Earlier concerns about large dams being socially and environmentally unsustainable and unethical make for a dramatic contrast with the current discourse of hydropower as a green and carbon-neutral climate mitigation strategy. 'Apocalyptic' imageries of a global climate crisis requiring urgent mediation (Swyngedouw 2010) facilitate the recasting of dams as the 'moral alternative' to fossil fuel energy (Fletcher 2010: 5). Yet, as Ahlers *et al.* (2015: 198) aptly note, 'repackaging hydropower infrastructure as clean energy is confusing the resource with the instrument: water is renewable, yet dams are not' (see also McCully 2001).

This global 'climate' push for hydropower has been complemented by a drive among newly dominant economies, like India, Brazil, Turkey and China, to accelerate their energy security strategising in order to maintain high rates of economic growth and modernisation. In many of these countries, fast-tracked dam development has been facilitated by the liberalisation of national energy sectors to enable private capital to boost the rate and speed of dam construction (Matthews 2012; Moore *et al.* 2010).

For the private investors, hydropower generation is a profitable business due to ostensibly cheap operating costs; in theory, the necessary fuel (water) is freely available. Moreover, the possibility to receive funding through climate finance and carbon markets is not only attractive economically, but also ensures political leverage (Erlewein and Nüsser 2011; Newell *et al.* 2011). The synergistic coupling of environmental and economic benefits characteristic of the twenty-first–century's neoliberal green governance paradigm (Goldman 2001) has served to conveniently position new hydropower developments outside the context of earlier dam-related critiques and concerns.

The aim of this chapter is to unravel the alleged synergies between hydropower, green growth and economic development by looking at some of the local controversies, contradictions and outcomes surrounding the uncritical promotion of hydropower by the State, political elites and private players. In India these dynamics emerged with the launch of the Government of India's 50,000 MW hydro initiative in 2003. The initiative envisioned the construction of 162 new hydropower projects by 2017, most of which are planned in the Himalayan region, the country's greatest repository of unexploited rivers (Dharmadikary 2008).

The majority of these projects will be developed by private, the so-called Independent Power Producers. Liberalisation of the national power sector in 2003 created a dizzying institutional landscape of public-private conglomerates, which significantly masks who is accountable for these interventions and their impacts (Erlewein 2012). The small Himalayan State of Sikkim (Map 5.1) was one of the forerunners in undersigning Memoranda of Understanding with private power developers and – with over 20 power projects in various stages of planning and implementation – counts among the top power producing Himalayan States.

Taking the case of Sikkim, this chapter points to some of the most controversial aspects of hydropower development in the Eastern Himalayas: its depoliticising promotion through a series of recurrent, one-sided narratives; the portrayal of 'run-of-the-river' hydropower technology as seemingly benign, despite the local evidence of severe environmental externalities for the fragile (till now well-conserved) Eastern Himalayan ecosystem (Menon and Vagholikar 2004; Kohli 2011); and the coercive repression of public debate and civil society contestation over the State's hydropower agenda (Schaefer 1995; see also Huber and Joshi 2015).

The study builds on fieldwork carried out in Sikkim between January and April 2011, and in January 2012, and is informed by two conceptual approaches. First, a political ecology approach, which allows us to map power configurations at multiple scales, and to understand

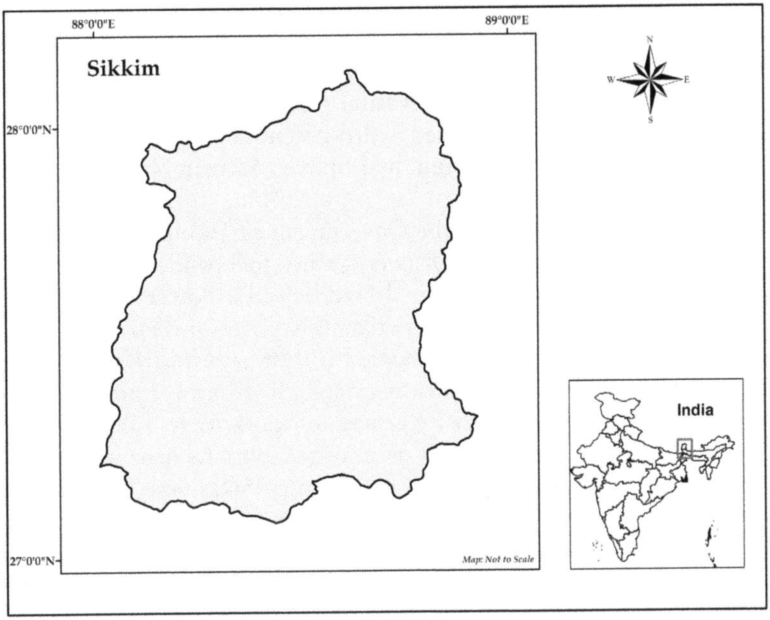

Map 5.1 Location of Sikkim in India

Source: SOPPECOM, Pune

how this shapes local responses and project outcomes (Baghel and Nüsser 2010). Further, political ecology scrutinises discursive and material political struggles and means through which dam-building is promoted, legitimised, obscured and/or negotiated.

The second approach illustrates how these new developments – especially new financial, technological and institutional modalities – rapidly reconfigure the institutional landscapes around water and energy at multiple levels, resulting in the loss of the 'political' in environmental and water governance (Swyngedouw 2009). Given the specificities of Himalayan hydropower, this chapter highlights the need to review official dam-related policies and regulatory frameworks, and to strengthen the space and scope for socio-environmental justice initiatives.

The politicised environment of hydropower development in Sikkim

Located along an important cross-Himalayan trade route, the Indian State of Sikkim was an independent monarchy until its annexation to the Indian Union in 1975. It is a geopolitically strategic border zone between Nepal, Bhutan, Tibet/China and West Bengal/India, and falls within the Eastern Himalayan biodiversity hotspot. The entirely mountainous terrain of the State is crisscrossed by multiple rivers and smaller streams and a steep elevation gradient (213 m to 8,598 m) endows Sikkim with an assessed hydro-potential of 8,000 MW,[2] making it the second most important hydropower State in Northeast India (Vagholikar and Das 2010).

Between 2001 and 2011, the Government of Sikkim signed Memoranda of Understanding with over twenty Independent Power Producers (including India's National Hydroelectric Power Corporation (NHPC)) for construction of more than thirty 'run-of-the-river' hydroelectric projects.[3] With Sikkim declared 100 per cent electrified in 1991, and its peak power demand met by existing hydropower projects,[4] the planned expansion of electricity generation capacity to 5,000 MW by 2015 is almost entirely geared at power export to urban-industrial agglomerations in other parts of the country. Sikkimese hydropower thus represents an important pillar of India's national energy security initiative. In order to legitimise this additional hydropower development, the Government of Sikkim takes recourse to several, rather one-sided pro-dam narratives.

A first narrative, expressed by the Chief Minister himself in popular metaphors of how the wealth held in Sikkim's rivers is 'washed away', and how, by damming the rivers this 'white gold' will be held back and harnessed locally, argues for the necessity to ensure the State's financial autonomy. However, according to data and statistics, Sikkim is the third richest economy among Indian States, with the third highest net state domestic product.[5] As a 'Special Category State',[6] Sikkim receives preferential funding from the Central Government for infrastructure development, health, education and poverty alleviation (Arora 2009), and the privilege of non-taxation until 2017 (The Sikkim Times 2011). There is little doubt that such nurturing will continue post 2017, given the strategic location bordering China.[7] Thus, the State government's positioning of hydropower as the only and best way to facilitate economic development in a State whose economy is largely rural, and which features hardly any industrial activity is somewhat contradictory.

Such narratives are supported by a gamut of promises about the expected pay-offs from hydro development: low variable costs of generation, employment, electrification, rural infrastructure and plentiful revenues. Impressive numbers are quoted: 15 billion Indian Rupees (Rs.) hydro earnings are expected to accrue annually over a total of

35 years (Energy & Power Department 2010), amounting to yearly per capita earnings of around 388 US$ – nearly a quarter of India's gross domestic product (GDP) per capita,[8] making Sikkim 'one of the richest states in the country' (Resneck 2010). By pledging to reinvest these royalties into the key State development objectives – infrastructural expansion, education, poverty eradication and employment generation – the Government of Sikkim in a 2009 publication anticipated that by 2015, Sikkim would be a prosperous, poverty-free State and 'a land of opportunities with zero unemployment' (GOS 2009: 12). Claiming that hydropower is the only realistic source of income to ensure economic growth, financial autonomy and socio-economic upliftment of the Sikkimese population thus turns the hydropower mission into a sort of moral imperative.

A second recurrent pro-dam discourse in Sikkim positions dams as being particularly beneficial to project-affected persons. Mandatory compensation requirements (as defined in national and State resettlement and rehabilitation (R&R) policy are presented locally as enormous individual gains: monetary payments for leased/acquired land; one job per fully project-affected household;[9] one per cent of the revenues generated for local area development; project-related employment according to skills, etc. (GOI 2007). Similarly, rural infrastructure development services (e.g. water supply, roads, community halls, and sanitation and waste facilities) that should be the responsibility of the welfare state – with or without hydropower development – are presented as additional sops in exchange for the sacrifices made by the affected communities. As one respondent aptly noted, such enticements are invariably effective:

> There might be opposition and scepticism regarding the project. But if the government opens a health dispensary and a playground before starting the project work, then all people will be in favour.

Several respondents whom we met in project affected areas also reported promises of unlimited, free-of-cost electricity – a benefit which has yet to materialise even in areas where hydropower projects are already commissioned (see Map 5.2 for the existing and planned hydroelectric projects in Sikkim).

The third narrative turns to environmental arguments and highlights the environmentally benign and low-impact nature of the run-of-the-river design chosen for new hydropower projects in Sikkim. While such claims are in themselves problematic, as will be discussed in the following section, they aim to reinforce the carefully constructed

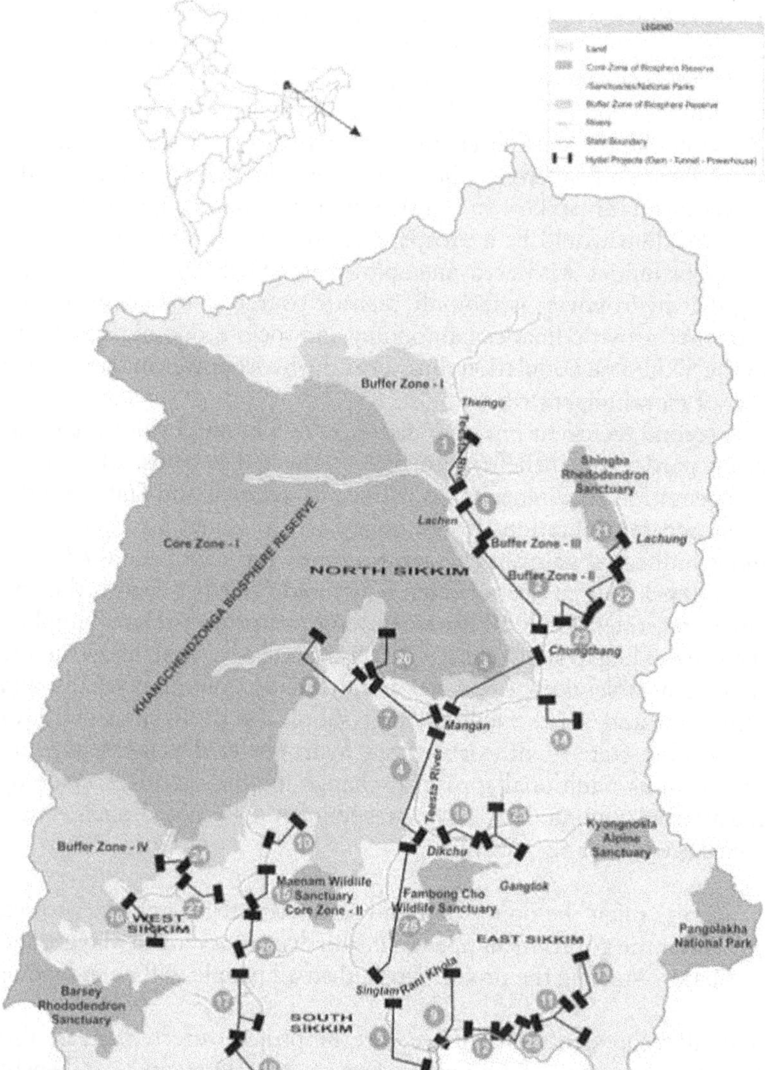

Map 5.2 Existing and planned hydroelectric projects in Sikkim. Roughly a third of the projects featured on this map have been shelved permanently or temporarily by 2015.

Source: Vagholikar and Das (2010)

image of an environmentally committed state leadership. Thus, during its 20-year rule, the governing party had bagged several national awards for investments in environmental protection measures, including large-scale afforestation and restrictions on tree-felling, a grazing-ban in high-altitude areas, the abolition of non-organic farming or a ban on plastic bags in the capital (*Down to Earth* 1999; GOS 2013).

That the Government of Sikkim's 'hydropower environmentalism' is more superficial than its earlier environmental commitments became clear when the same state leadership reacted aggressively to initial concerns raised by environmentalists about the inappropriate environmental impact assessments for dams in the region. Concerns relating to speedy and uncritical green clearance approvals of projects located in an ecologically and geologically fragile, seismically active and culturally unique area (Vagholikar and Das 2010) were dismissed by the Chief Minister's office as dissent masterminded by anti-national, anti-development outsiders, to defame the peaceful, conflict-free democracy of Sikkim.

Such discursive distortions also reflect the coercive political culture, which has enduringly stemmed the emergence of hydro-critical voices and resistance to dams in Sikkim, as we discuss in the last section of this chapter. Statements such as the following, made by the Chairperson of the Sikkim Pollution Control Board at the public hearing for the 1,200 MW Teesta III HEP in 2006, warn of the consequences of voicing alternative opinions and give a hint that the carefully staged image of a democratically vibrant state is more reminiscent of a democratic façade:

> You *should* reap the benefit because no one can stop this project, no matter which political party comes to power tomorrow. No one can stop this as the Government of India has given the *orders* . . . These projects are not meant to harm or bring tension to anyone . . . Anyone who disturbs this project is not a Sikkimese. He might be born in Sikkim but is a *useless person* if he opposes such a good project; such people are your opposition and *anti-social elements* . . . Since you are opposing the Government of India you are an *anti-national* . . .
>
> (Save Dzongu n.d.)

Run-of-the-river: controversial dam design and socio-environmental implications

A lot has been written about the controversial labelling of large-scale run-of-the-river (R-o-R) hydropower projects as socially and environmentally benign (see for example Kohli 2011; Vagholikar and Das 2010). No doubt, the R-o-R design is viable for small-scale schemes on mountain streams. Such schemes have existed in Sikkim since 1927, traditionally operating with above-ground water channels or large pipes, and have a limited impact on the river and mountain ecology. However, the currently proposed R-o-R schemes are of much larger dimensions, and require large dam structures and extensive underground tunnel networks. In a comparatively small state like Sikkim, where arable land is a scarce resource, acquisition of agricultural land can disproportionately impact local populations (Figure 5.1).

Moreover, while physical displacement due to submergence or land acquisition is indeed minimal in Sikkim's R-o-R schemes, compared to conventional storage dams, there are a number of indirect impacts on communities not physically dislocated, which are not accounted for in existing R&R policy. The several kilometre-long water diversion tunnels, for example, are excavated through rock-blasting, using explosives. This adversely affects the relatively young, geo-ecologically fragile Himalayan landscape, its geo-hydrology and the

Figure 5.1 Teesta disappearing into diversion tunnel at 1,200 MW Teesta III HEP in Chungthang, North Sikkim

Source: Photo: Amelie Huber

rural communities settled on these mountain slopes (see also Menon and Vagholikar 2004; Bhutia 2012). Thus, primary research in the 510 MW Teesta V project area indicated significant social and economic consequences from the short and long-term impacts of R-o-R projects. Numerous experiences of gradual and sudden land movements were reported, especially in locations around the tunnel networks, resulting in damage to homes and agricultural fields. In some areas the residents have chosen to migrate due to fears of more disastrous consequences.

In September 2011, a 6.9 Richter earthquake hit Sikkim. Although data and statistics are missing, several individuals observed earthquake-induced damages and destructive landslides to be more acute in the vicinity of hydropower construction sites, compared to other areas with similar geological conditions. Both locals and experts assume that this was due to slope destabilisation through tunnel blasting (Kohli 2011; Manish 2011). Furthermore, affected communities across Sikkim are reporting that natural springs (the main source of water for agricultural and domestic use) have dried up in areas located near the tunnels. That blasting and tunnel construction disturb sub-surface hydrogeological conditions has also been confirmed as plausible by a number of scientific experts, including the State Department of Mines, Minerals and Geology (see Bhutia 2012).

We flag the above issues to reiterate the fact that there is no accounting for these localised impacts of R-o-R schemes in national policy frameworks for environmental management and rehabilitation. For example, the national Resettlement and Rehabilitation policy (GOI 2007), being largely designed for non-mountain contexts, targets solely those affected by land acquisition as project-affected people (PAP). In the mountains of Sikkim however, R-o-R projects have physically, economically and perhaps psychologically affected much larger numbers of rural inhabitants, and have also led to significant displacement post-project construction – the Teesta V HEP being a real-time case in point.

The potentially hazardous relationship between hydropower projects and fragile mountain ecologies was acknowledged officially at the national level after June 2013, when a catastrophic flash flood killed an estimated 1,000 people in the Central Himalayan State Uttarakhand, another hotbed of hydropower development. This event and the recent earthquakes in Sikkim (2011) and Nepal (2015) have heightened local fears around dam projects in mountain regions. In the wake of the flash flood disaster, India's Supreme Court directed the MoEF to halt environmental clearance for further hydroelectric power projects in Uttarakhand, and to constitute an expert body that would prepare a detailed study on whether hydroelectric projects contributed to the massive environmental disaster (The Indian Express 2013). In Sikkim in turn, the incident had no repercussions. Instead, in order to facilitate the clearance of several contentious hydropower projects, the Government of Sikkim backed a policy amendment by the MoEF to reduce the buffer zones around Sikkim's national parks and wildlife sanctuaries from 10 km to a mere 25–200 m (Shrivastava 2014).

The apolitical construction of popular consent

In the remainder of this chapter our story focusses on the local political repercussions of these dam projects and their uncritical, apolitical promotion. Here it is interesting to note that the State government's rushed allocation of several dozen projects to mostly private power producers under Build-Own-Operate-Transfer arrangements had generated comparatively little critique locally. Contestation of the dam projects by project-affected communities and civil society groups took significant time to emerge. We show that the initially disparate citizen reactions to dam projects in Sikkim were the outcome of a constraining political landscape that significantly curbed dissent. However, we argue that the State-imposed hydropower agenda eventually triggered a contrary repoliticisation of state–society relations, as conflicts over hydropower exposed the repressive tendencies of the state apparatus, and the fact that democracy is a rather ill-defined concept in Sikkim (Arora 2008). This triggered an unexpected public rebellion of previously submissive communities and individuals against the coercive hydropower agenda, and the coercive state itself.

Sikkim's patronage-based system of doing politics has its roots in the State's political history, which has long been characterised by an ethnicity-based mode of governance (Chettri 2013). The electorate of Sikkim is divided (somewhat crudely, omitting a range of long-term residents) into the following main ethnic categories: the indigenous Lepchas and Limbus (Tsongs) – the smallest minorities; the economically and politically more astute Bhutias who migrated from Tibet, constituting the old Monarchy and landed elite; and the Nepali majority, dominating State politics today, but by no means an ethnically homogenous entity. From the days of the old monarchy to the 1975 annexation to India, ethnic fractures have persisted in Sikkim and have been reiterated and instrumentalised by political parties (Chakraborty

2000; Chettri 2013; Schaefer 1995). While Sikkim is now an Indian federal State, a 'Sikkimese'/'non-Sikkimese' divide defines the complex inclusion and exclusion of residents in Sikkim:

> Some may sell land to whomever they please, others may not. Some can enter all of Sikkim; others need special permits for certain areas. Some hold Nepali citizenship, some may not marry a non-Sikkimese and keep taxation-related and social and financial benefits . . . Within Sikkim, groups position themselves as insiders in order to gain political leverage, their insider status strengthened by their tribal tenure on the land. The insider/outsider frame is a common measure of someone's identity and rights in Sikkim.
>
> (Little 2010: 115, 116)

A constructed consensus is easier to achieve in politics by making some people more equal than others through 'differential and unequal social and political citizenship rights' (Swyngedouw 2011: 3). This is precisely how political power has been used to divide and rule in Sikkim. For large parts of the rural population, patronage and government favouritism still represent the economic basis of livelihood provision. Thus while the State, given its strategic geographical position, receives preferential funding from the Central Government of India, it has failed to sufficiently and equitably promote entrepreneurship and livelihood autonomy among citizens (Schaefer 1995).

The normative function of all governments is to deliver development goods and welfare services. In Sikkim, however, governments have always used this function to establish the notion of the State as a generous benefactor. This is done through bias in the extension of livelihood aid, such as building materials, seeds, livestock, food rations, water supply and sanitation facilities, and even entire model brick-homes (all in the name of poverty alleviation) to 'conforming' individuals, households or political administrative units. Moreover, since the Government of Sikkim is the largest employer in a State where youth unemployment is rampant, access to popular government jobs, business contracts or professional licenses is equally regulated through patronage relations.

In such a setting, going against the grain and voicing contrary opinions in public holds a high risk of 'victimisation'. This term is widely used in Sikkim to denote a much dreaded process of 'othering' and punishing those who criticise or oppose the State government. It works either discursively, through a distinct ostracising narrative, or materially, by withholding government patronage. Since the State is small, the likelihood of being watched or overheard is significant, and political pressure can easily be extended across the relatively short chain of politicians, bureaucrats, middlemen, party-associates and village leaders, thereby effectively curbing critical, antagonistic state–society relations:

> The government employs practically every educated soul, none of whom dares express his opinion for fear of losing his sinecure. As a result there are no independent thinkers, no intellectual circles, and only a few isolated individuals capable of openly analysing and expressing a critical opinion for the benefit of the state's development.
>
> (Schaefer 1995: 14)

Sikkim's fractured ethnic fabric and its coercive politics explain to some extent why the current Chief Minister (CM) Pawan Chamling could enter his twentieth consecutive year as head of State in 2014. This is an atypical situation in a State where, in the Chief Minister's own words, 'democratic decorum prevails uninterrupted' (Sikkim Now! 2012). Many critics would argue that Chamling's government has persisted due to a successful decimation (read victimisation) of opposing persons and parties.

The strong ethnic divides in the State are also partly the reason why local responses to dams have been rather disparate and spatially confined, causing many to label Sikkim's most vocal instance of community dissent to dam-building – the successful anti-dam campaign staged by the Affected Citizens of Teesta – as a 'Lepcha thing' (Little 2010: 121). Indeed, what has been the most fervent anti-dam struggle in the State to date was articulated by a group of predominantly young Lepcha activists, who staged several rounds of vociferous protest against six projects planned within the protected tribal Lepcha reserve Dzongu in North Sikkim. One of the key factors that triggered this conflict was the fact that entry to this culturally and environmentally significant area (considered by many Lepchas as their 'sacred homeland'; Little 2010: 109) had historically required a complex set of administrative approvals, except for the indigenous Lepcha community living in the region. The fact that almost overnight these restrictions were overruled, and the area was sanctioned to become a major dam construction site was an ironic and unpalatable shock to many Lepchas within and beyond Sikkim.

However, an in-depth analysis of the muted response to the dam projects in other parts of Sikkim shows more complex political reasons, in other words a coercive political culture, which reflects a significant democratic deficit. The fact that patronage relations have been a major factor in subduing popular dissent to proposed hydropower projects is reflected in the following statements by members of project-affected communities (interviewed between February and April 2011):[10]

Will you protest the project?
If the project comes and all people accept it, how can I protest? The public cannot protest the government.

Why is that so?
People are afraid of the government; they are totally dependent. If people protest they can't get the benefits the government is giving out. The majority is against the project but no one will be openly opposing. A knowledgeable person said we should not complain.

Who says these things?
Village leaders, party members . . . one month ago the Chief Minister came to give a speech, saying that project construction will start, and we should let it go ahead.

The interviews also highlighted how 'high levels of corruption which plague [. . .] administration and political parties' remain unquestioned (Schaefer 1995: 14). The local panchayat of a project-affected village said:

> We have no power, that's the problem. The government is imposing this project and there is a lot of corruption. The government has its channels, so we can't avoid this.

The fear of being labelled 'anti-social', 'anti-national' or of being deprived of benefits is particularly widespread among those who currently 'enjoy' such favours. This includes, among others, public servants, grassroots politicians and even retirees, who depend on government pensions and subsidies. The following statement was made by a teacher, one of the most vulnerable professional groups, as the State reserves the right to relocate them at will to any part of the state:

> At the public hearing I opposed the project and spoke out, but what to do for the simple public? We don't have any opportunity. I protested three times. The last time I finally agreed, because of political pressure. Many people here are employed by the government and exist only through its benevolence.

Political pressure can also affect entire villages where villagers jointly defend their cause. As one female panchayat leader related:

> The community is against the project. The panchayat should listen to the political leaders but I am with the community. We know we will get lots of problems – they will likely stop providing some basic services.

Also, some private project developers took advantage of this skewed patron-client system. As villagers related, 'even before the project the company people came and visited individual homes, gave presents, etc. They gave benefits for the villagers and participated in village functions, such as weddings and funerals'. Such buy-in tactics allow for mandatory information to be selectively presented, for instance in hushed discussions with landowners on compensation for land acquisition. Similarly, while project benefits tend to be over-publicised, information on negative externalities are often diluted and withheld. According to legal requirements, each public hearing on a particular project must be sufficiently advertised, documented, and the report submitted to the national Ministry of Environment and Forests for review and inspection. Yet many respondents spoke of not having been informed adequately or in time; of having to travel for several hours to the hearing location and back – travel expenses to be self-borne; and of not being able to effectively express their concerns at such gatherings for fear of victimisation.

Lack of a transparent public space, so vital to democracy, is evident at every governance level in Sikkim. This explains what Swyngedouw identifies as the difference between 'politics' and the 'political' – a political façade:

> *Politics* is the power plays between political actors and the everyday choreographies of policy making, and its institutional-procedural configurations, in which individuals and groups pursue their interests'. To the contrary, 'a foundational gesture of a political democracy is [an] empty space of power. [. . .] It is always disruptive, it emerges as the refusal to observe the 'place' allocated to (particular) people or things.
>
> (Swyngedouw 2011: 3–7)

The carefully constructed, 'peaceful democracy' in Sikkim mean that the State, its leadership and its decisions reign supreme:

> We are poor farmers, depending on the government. We are just standing by as the government goes ahead with its plans. There is

> no interaction with the community . . . It's all up to the government, they have a free hand.
>
> . . . Even though they went on hunger strike [in Dzongu] nothing happened. It all depends on the government, they are the all-powerful. The protesters can't really influence the government nor can the people of the village. Even the landowners are just guardians of the land; the real owner of the land is the government.

And yet, as stifling as the coercion and repression of public debate over hydropower in Sikkim may seem, the civil resistance that emerged sowed the seeds for an unexpected citizen engagement with the development politics of the State; and for a slow rupture with the submissive behaviour when confronting the government, so characteristic to state–society relations in Sikkim. When State-imposed dam construction violated the deep-rooted cultural and religious spaces and sentiments of indigenous communities, the 'sutures' of patronage and coercion were broken (Swyngedouw 2011: 4), galvanising an uncompromising posture of ACT activists, who declared they 'will accept nothing less than a complete scrapping of hydel projects' in spaces culturally sanctified to them (Arora 2008: 27). The State government reacted by branding Lepcha activists 'anti-social', 'anti-national' elements, threatening them and their family members' with exclusion from public opportunities, public offices, or transfers to 'punishment postings'. Indeed some of these coercive tactics worked and many switched sides.

However, despite the continued coercion, critical local voices are increasingly being voiced and heard, within and outside of Sikkim (Little 2010). The hydropower mission that is being advanced at a rapid pace, helped by a weak regulatory framework, seems to have pushed affected citizens to reclaim a democratic-political space. There are subtle, maybe premature indicators that this re-politicisation of environmental governance – and the governance of rural development more generally – links historic ethnic divides, and helps unite localised environmental justice struggles, as evident in the novel coalition of advocacy groups 'The Common Platform for Joint Action against Hydropower Projects' in late 2011 (Sikkim Now! 2011).

Another example of a 'properly political' backlash against the increasingly coercive state apparatus occurred in 2011, when the controversial 'Sikkim Prevention and Control of Disturbance of Public Order Bill' was tabled by the State government in order to contain dissent. This bill, which banned processions, hunger strikes, squatting, sloganeering and other forms of public agitation, was termed as the 'black bill' by opposition parties. Contrary to government expectations, the tabling of this bill caused a State-wide uproar that eventually prompted a humiliating recall by the State leadership (The Telegraph 2011).

In 2013 the demand for more profound political-democratic change became further manifest in the rapidly growing public support for the 'Sikkim Revolutionary Front' (SKM; Talk Sikkim 2013). In an unprecedented feat this new political party garnered 40.8 per cent of votes in the 2014 State Legislative Assembly elections, managing to wrest ten seats from Chamling's ruling Sikkim Democratic Front (SDF).

Conclusion

Hydropower development is one among a range of new neoliberal environmental projects propagated worldwide with a consistently uncritical approach. Our findings presented in this chapter suggest an urgent need to unpack sustainable hydropower discourse, and to expose its endeavour to greenwash what in reality are large-scale, often socially and ecologically disruptive interventions. Contemporary hydropower propaganda tends to be completely detached from earlier criticisms and concerns over large dams, and conceals the serious environmental, social and political problems with which hydropower development continues to be fraught.

The divergent responses to hydropower in Sikkim and across the Eastern Himalayas reflect the serious lacunae in how information about hydropower infrastructure is diffused. This being a geographic area where a majority of the population has no access to balanced information, the social and political responsibility of decision-makers to ensure free, prior and informed consent becomes more acute. As our findings suggest, in Sikkim the administration has failed to live up to this responsibility. What is more, it has sought to actively intimidate and suppress those voices that have publicly flagged these shortcomings. In doing so they grossly underestimated the power of civil society dissent, and how the failure to address serious public concerns can provide the tipping point for change, in this case heightening citizen awareness for environmental justice and democracy.

Sikkim has experienced a remarkable people's movement, which has effectively challenged the State. Sadly, these achievements have not been widely recognised in the State itself. We therefore see an acute need for research, political and social activism to give visibility and support to such nascent citizen struggles, since they represent a much-needed political control mechanism, which so far has been lacking in this and similar superficially democratic contexts. We believe that a first step to encourage citizen engagement with development politics is to recognise the successes and learning experiences, however small they may be, of these social and environmental justice initiatives.

The need for electrification, energy security and economic development in certain parts of India no doubt needs to be taken seriously. Hydropower is but one source of energy that can be harnessed for these different purposes, and to do so there are different technical options and designs, including options that tend to be much less publicised, such as micro, mini and small hydropower. These must be included in energy security debates. Our chapter has provided just a snapshot of some of the difficulties associated with planning and construction of large-scale hydraulic infrastructure in a geologically and ecologically highly complex and fragile terrain. Nevertheless, the findings are unambiguous: current national regulatory frameworks appear ill-designed to address the impacts of large run-of-the-river dams on the Eastern Himalayan mountain ecosystems, or to regulate the neat connivance among a powerful triad of national, state and private hydropower promoters.

This chapter underlines the urgent need to modify existing social and environmental safeguard policy (including Environmental and Social Impact Assessments, as well as Rehabilitation and Resettlement Policy), in order to take into account these complexities common to a number of mountain States and regions. In an effort to avoid increasingly adverse and unsafe impacts of hydropower development on local societies and environments, this would no doubt imply in certain cases opting for a different site, technology, project size or even to refrain from infrastructure development altogether. It may be obvious, but worthwhile to recall anyways, that hydropower is no easy one-size-fits-all solution.

Notes

1 A substantially different version of the chapter by the authors titled 'Hydropower, Anti-Politics, and the Opening of New Political Spaces in the Eastern Himalayas' is available in *World Development* (2015), pp. 13–25.

2 www.sikkimpower.org/power/about_us.aspx.

3 Two more projects – Rangit III and Teesta V – had been allotted to NHPC before 2000. At least six more projects had been planned but were subsequently cancelled by the Government of Sikkim. Of the remaining projects six have been cancelled only recently, in 2012. (Energy & Power Department, Government of Sikkim www.sikkimpower.org/power/files/Status_of_HEPS.pdf.)

4 Sikkim's current peak power demand has been met after completion of the 510 MW Teesta V hydroelectric project (HEP) in 2008 (Energy & Power Department 2010).

5 The Net State Domestic Product (NSDP) is a measure to define, in monetary terms, the volume of all goods and services produced within the boundaries of the state during a given period of time. See Ministry of Statistics and Programme Implementation (http://pib.nic.in/newsite/PrintRelease.aspx?relid=123563).
6 Special Category Status has been awarded to eleven Indian border states (Arunachal Pradesh, Assam, Himachal Pradesh, Jammu and Kashmir, Manipur, Meghalaya, Mizoram, Nagaland, Sikkim, Tripura and Uttarakhand) with harsh/hilly terrain, sparse and significant tribal population and a low level of infrastructural development, with consequentially high delivery cost of public services, backwardness and social problems. Special Category States receive preferential treatment in federal assistance as well as tax breaks to fuel investment in industrial development (Arora 2009; Saxena 1999).
7 While there have been allegations about non-continuation of development funding from the centre, it is unclear to what extent these can be taken seriously. Since there are still disputed territories along the border with Tibet/China, it is in the interest of the Central Government (which has a strong army presence in Sikkim) that internal peace and tranquillity, and overall citizen satisfaction are maintained in the state (Little 2010).
8 Exchange rate as per 31.12.2014, population as per 2011 (http://censusindia.gov.in/), GDP as per 2014 (http://data.worldbank.org/indicator/NY.GDP.PCAP.CD).
9 Officially, two categories of project-affected people (PAP) are recognized: Fully PAP whose entire property has been acquired, and partially PAP whose land has been partially acquired.
10 The following are statements by members of different project-affected communities (Sikkim February–April 2011).

References

Ahlers, R., Budds, J., Joshi, D., Mehta, S., Merme, V., and Zwarteveen, M. 2015. Framing hydropower as green energy: Assessing drivers, risks and tensions in the Eastern Himalayas. *Earth Systems Dynamics*, 6: 195–204.

Arora, V. 2008. Gandhigiri in Sikkim. *Economic and Political Weekly*. 20 September, pp. 27–28.

Arora, V. 2009. They are all set to Dam(n) our future: Contested development through hydel power in democratic Sikkim. *Sociological Bulletin*, 58(1): 94–114.

Baghel, R. and M. Nüsser. 2010. Discussing large dams in Asia after the World Commission on Dams: Is a political ecology approach the way forward? *Water Alternatives*, 3(2): 231–248.

Bhutia, D. T. (ed.). 2012. *Independent People's Tribunal on Dams, Environment and Displacement*. New Delhi: Human Rights Law Network. http://hrln.org/hrln/publications/books/930-independent-peoples-tribunal-on-dams-environment-and-displacement.html (Accessed on 5 January 2016).

Chakraborty, J. 2000. Sikkim: Elections and Casteist Politics. *Economic and Political Weekly*, 28, October, pp. 3805–3807.

Chettri, M. 2013. Ethnic Politics in the Nepali Public Sphere: Three Cases from the Eastern Himalaya (PhD thesis). SOAS, University of London.

Dharmadikary, S. 2008. *Mountains of Concrete: Dam Building in the Himalayas*. Berkeley, CA: International Rivers. www.internationalrivers.org/files/attached-files/ir_himalayas.pdf (Accessed on 5 January 2016).

Down to Earth. 1999. Taking the lead, 7(18). February 15, 1999. www.downtoearth.org.in/coverage/taking-the-lead-19340 (Accessed on 5 January 2016).

Energy & Power Department. 2010. *Energy & Power Sector Vision 2015*. Gangtok: Energy and Power Department, Government of Sikkim.

Erlewein, A. 2012. Energie aus dem Himalaya: Ursachen und Folgen des Wasserkraftbooms in Himachal Pradesh, Indien. *Geographische Rundschau*, 64(4): 25–33.

Erlewein, A. and M. Nüsser. 2011. Offsetting greenhouse gas emissions in the Himalaya? Clean development dams in Himachal Pradesh, India. *Mountain Research and Development*, 31(4): 293–304.

Fletcher, R. 2010. When environmental issues collide: Climate change and the shifting political ecology of hydroelectric power. *Peace and Conflict Review*, 5(1): 1–15.

GOI (Government of India). 2007. *The National Rehabilitation and Resettlement Policy, 2007*. New Delhi: Government of India, Ministry of Rural Development.

Goldman, M. 2001. Constructing an environmental state: Eco-governmentality and other transnational practices of a 'green' World Bank. *Social Problems*, 48(4): 499–523.

GOS (Government of Sikkim). 2009. *Sikkim – 15 Triumphant Years of Democracy (1994–2009): A Call to Action*. Gangtok: Government of Sikkim, Information and Public Relations Department. http://smilingsikkim.org/wp-content/uploads/2010/12/Part_1.pdf (Accessed on 5 January 2016).

GOS (Government of Sikkim). 2013. *Achievements, Recognitions and Awards*. Gangtok: Government of Sikkim Website. www.sikkim.gov.in/portal/portal/StatePortal/Awards/AchievementsContent (Accessed on 5 January 2016).

Huber, A. and D. Joshi. 2015. Hydropower, anti-politics, and the opening of new political spaces in the Eastern Himalayas. *World Development*, 76, 13–25.

The Indian Express. 2013. SC shackles govt: No new hydro power project in Uttarakhand for now. *The Indian Express*. 14 August 2013. www.indianexpress.com/news/sc-shackles-govt-no-new-hydro-power-project-in-uttarakhand-for-now/1155178/ (Accessed on 5 January 2016).

Khagram, S. 2004. *Dams and Development: Transnational Struggles for Water and Power*. New York, NY: Cornell University Press.

Kohli, K. 2011. Inducing vulnerabilities in a fragile landscape. *Economic and Political Weekly*, 46(51): 19–22.

Little, K. 2010. Democracy reigns supreme in Sikkim? A long march and a short visit strains democracy for Lepcha marchers in Sikkim. *Australian Humanities Review*, 48: 109–129.

Manish, S. 2011. A paradise dammed. *Tehelka*, 8(40), 8 October 2011. http://archive.tehelka.com/story_main50.asp?filename=Ne081011PARADISE.asp (Accessed on 5 January 2016).

Matthews, N. 2012. Water grabbing in the Mekong basin – an analysis of the winners and losers of Thailand's hydropower development in Lao PDR. *Water Alternatives*, 5(2): 392–411.

McCully, P. 2001. *Silenced Rivers – The Ecology and Politics of Large Dams*. Enlarged and Updated Edition. London: Zed Books.

Menon, M. and N. Vagholikar. 2004. *Environmental and Social Impacts of Teesta V Hydroelectric Project, Sikkim: An Investigation Report*. New Delhi: Kalpavriksh Environmental Action Group.

Moore, D., J. Dore and D. Gyawali. 2010. The world commission on dams + 10: Revisiting the large dam controversy. *Water Alternatives*, 3(2): 3–13.

Newell, P., J. Phillips and P. Purohit. 2011. The political economy of clean development in India: CDM and beyond. *IDS Bulletin*, 42(3): 89–96.

Resneck, J. 2010. The Green Mountain State. 1 December 2010. www.caravanmagazine.in/journeys/green-mountain-state (Accessed on 5 January 2016).

Save Dzongu. n.d. *The Background*. http://savedzongu.wordpress.com/background/ (Accessed on 5 January 2016).

Saxena, N. C. 1999. *Medium-Term Fiscal Reforms Strategy for States*. Draft Discussion Paper, Planning Commission, Government of India. http://planningcommission.nic.in/reports/articles/ncsxna/index.php?repts=fiscal.htm#V.%20Special (Accessed on 5 January 2016).

Schaefer, L. 1995. A Sikkim awakening. *Himal South Asian*. September/October 1995. http://old.himalmag.com/component/content/article/2934-A-Sikkim-awakening.html (Accessed on 5 January 2016).

Shrivastava, K. S. 2014. Ministry in a hurry. *Down to Earth Magazine*. April 15, 2014. www.downtoearth.org.in/coverage/ministry-in-a-hurry-43879 (Accessed on 5 January 2016).

Sikkim Now! 2011. Groups protesting HEPs on Rathong Chu consider mounting larger agitation. *Sikkim Now*! 13 December 2011. http://sikkimnow.blogspot.de/2011/12/groups-protesting-heps-on-rathong-chu.html?q=common+platform (Accessed on 5 January 2016).

Sikkim Now! 2012. State Day message of Chief Minister Pawan Chamling. *Sikkim Now*! 18 May 2012. http://sikkimnow.blogspot.it/2012/05/state-daymessage-of-chief-minister.html (Accessed on 5 January 2016).

The Sikkim Times. 2011. Sikkim's tax free zone status to expire in 2017. *The Sikkim Times*. 15 August 2011. http://sikkimnews.blogspot.de/2011/08/sikkims-tax-free-zone-status-to-expire.html (Accessed on 5 January 2016).

Swyngedouw, E. 2009. The antinomies of the postpolitical city: In search of a democratic politics of environmental production. *International Journal of Urban and Regional Research*, 33(3): 601–620.

Swyngedouw, E. 2010. Apocalypse forever?: Post-political populism and the spectre of climate. *Theory Culture Society*, 27: 213–232.

Swyngedouw, E. 2011. Interrogating post-democratization: Reclaiming egalitarian political spaces. *Political Geography*, 30(7): 370–380.

Talk Sikkim. 2013. Mass effect: PS Golay creates history as huge numbers turn out in the capital to support him. *Talk Sikkim*, October 2013, pp. 16–24.

The Telegraph. 2011. Sikkim bill for public order. *The Telegraph (Calcutta, India)*. www.telegraphindia.com/1110812/jsp/frontpage/story_14369595.jsp (Accessed on 5 January 2016).

Vagholikar, N. and P. J. Das. 2010. *Juggernaut of Hydropower Projects Threatens Social and Environmental Security of Region*. Pune: Kalpavriksh; Guwahati: Aaranyak and New Delhi: ActionAid. https://chimalaya.files.wordpress.com/2010/12/damming-northeast-india-final.pdf (Accessed on 5 January 2016).

WCD [World Commission on Dams]. 2000. *Dams and Development: A New Framework for Decision-Making*. The Report of the World Commission on Dams. London: Earthscan. www.unep.org/dams/WCD/report/WCD_DAMS%20report.pdf (Accessed on 5 January 2016).

6 State Water Policy of Assam 2007

Conflict over commercialising water

Chandan Kumar Sharma

After the draft of the new State Water Policy 2007 was thrown open for public consultations in Assam in September that year, the policy immediately became a centre of serious controversy with peasant groups, civil society organisations, intelligentsia and media which vehemently opposed the draft policy. They alleged that the provisions in the latter would lead to commercialisation and privatisation of water besides taking away the traditional rights of the communities on water resources jeopardising their life and livelihood.

Assam (Map 6.1) is a predominantly agrarian economy. Its agrarian rural population including the tribal communities traditionally depend on the natural water bodies like the rivers, ponds, *beels* (wetland), streams, rivulets, etc. which are spread all over the State for free natural water for agriculture and other domestic uses. Besides, communities such as the fisher folk (e.g. the *kaibarta* community) are also directly dependent on these water bodies for their livelihood. Historically, these communities had free access to the water bodies as commons which played a critical role in fulfilling their subsistence needs.

Onslaught on such commons including their commercialisation (leasing out to *mahaldar*s) and encroachment by powerful business interest have already caused tremendous hardship and disgruntlement among the poor and marginal communities of the State. Simultaneously, the unbridled exploitation of groundwater in the urban areas by private parties (commercial or domestic) has also become a matter of serious concern. For example, in Guwahati, the capital city of the State, such exploitation has already caused fast depletion of its groundwater level. Such a crisis is looming large over other cities of the State too.

Genesis of the present conflict

In the context of the above, some regulation and management of water resources appears well in order, and from that point of view,

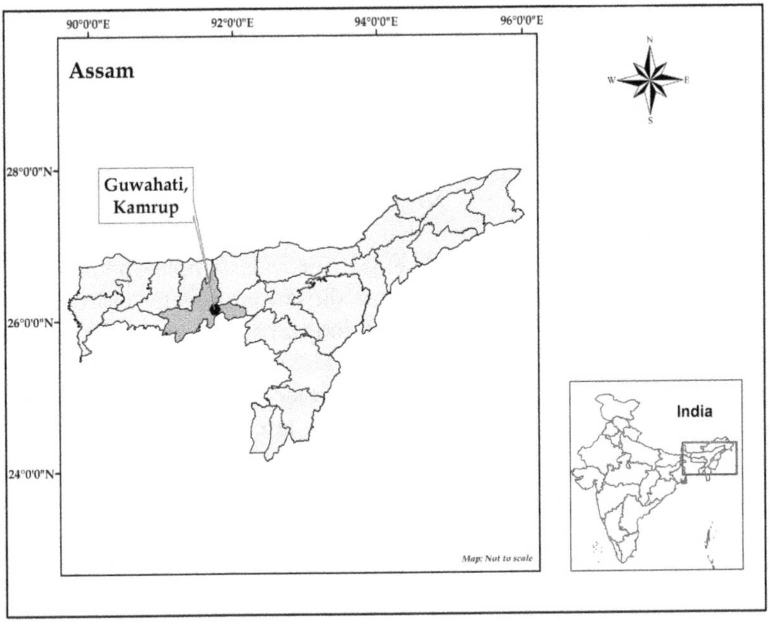

Map 6.1 Location of Assam and the State capital Guwahati
Source: SOPPECOM, Pune

the attempt at bringing about a water policy for the State of Assam seemed a welcome step on the part of the Government of Assam. In fact it was imperative under the National Water Policy 2002 that the States of the country would develop their water policies within a stipulated time frame. The National Water Policy (NWP 2002) said that 'the success of the National Water Policy will depend entirely on evolving and maintaining a national consensus and commitment to its underlying principles and objectives. To achieve the desired objectives, State Water Policy backed with an operational action plan shall be formulated in a time bound manner say in two years' (MoWR 2002).

However, the Draft State Water Policy 2007, far from meeting the general expectation of the people, raised many questions and concerns regarding its intent and potential implication.

The DSWP was prepared by the Assam Science Technology and Environmental Council (ASTEC) on behalf of the Water Resources Department of the Government of Assam. The draft was examined by a Task Force on State Water Policy constituted by the government in a meeting in August 2007. The Task Force was constituted by experts, present and former senior government officials and two representatives from civil society organisations. In that meeting, an opinion was expressed for wider public consultation of the draft.

Subsequently, in a press meet in Guwahati on 8 September 2007, the North East Chapter of Jalbiradri, a nongovernmental organisation (NGO), announced that they would hold public consultations across the State to take people's views on the DSWP. In the same press meet, Senior Gandhian and North East Chapter Chairman of Jalbiradri, Natwar Thakkar stated that the new DSWP bears clear signal of commoditisation of water. The draft provides ample opportunities for accelerating the process of commoditisation of water resources of the State by facilitating the participation of private corporate players in their exploitation and distribution. He, therefore, held that if the people of the State did not raise their voice against the draft policy, they would have to pay a very heavy price later on.

Points of disagreement

Accordingly, with the initiative of NE Jalbiradri, nine consultations were held across Assam spanning over a period of almost one month starting from 9 September 2007, in which the DSWP in its given form was rejected outright. It not only pushed people onto the streets in protest, but also prompted many academics, social activists and journalists to air their views through the media, protest meetings and public consultations on the DSWP.

The civil society in Assam participated in the debates and discussions on the DSWP with considerable amount of interest and vigour. The intelligentsia and the Assamese language media also took a critical stand on the DSWP. Their voices lent ideological and intellectual direction to the protests against the DSWP since the time the draft was announced in August 2007. The protests were spearheaded mainly by the popular peasant organisation, Krishak Mukti Sangram Samity (KMSS). Besides Jalbiradri, several NGOs from different parts of the State such as Rural Volunteers Centre, People's Movement for the Subansiri and Brahmaputra Valley, North East Social Research Centre, Church's Auxiliary for Social Action, etc. were involved in the organisation of public consultations and in the preparation of a critique of the DSWP. The framework for an alternative water policy for the State emerged from these consultations and critical observations.

The overwhelming popular temper expressed regarding the DSWP can be described as one of disappointment and resentment. In one of the early observations on the original draft of the DSWP, a former Director of Soil Conservation of the Government of Assam maintained that water was a public property which should not be altered to facilitate individual usage or profit making. He criticised the draft for not giving proper attention to watershed development for a sustainable and balanced water cycle (Dutta 2007). Another social activist claimed that the DSWP was nothing but a conspiracy to privatise water resources under the diktat of the World Bank and International Monetary Fund. The draft, he alleged, clearly facilitated privatisation, river linking and construction of big dams to the detriment of the interest of the common masses. Though the draft spoke of cooperation with local communities while implementing it, it would only pave the way for the entry of private capital through the backdoor (Sharma 2007). An eminent geology Professor observed that the DSWP was an outcome of the Union Government's pressure on the State governments to fall in line with the National Water Policy 2002 which authorised the Union Government to centralise and privatise water resources besides undertaking projects for interbasin transfer of water (IBTW) within the country (popularly known as the river linking project). Despite water resources being a State subject according to the Indian Constitution, the NWP 2002 paved the way for considering the water resources of the country as national assets and its planning, development and management in a national perspective (Goswami 2007). He asserted that in the NWP considerable emphasis has been laid on ascertaining the participation of the private sector in the projects related to water resources. He pointed out that in order to ensure that the common people were not subjected to unbridled exploitation by the privatisation of water resources, the National Water Advisory Committee (NWAC) suggested to the Union government that 'since all water resources have a common property character, private sector participation in planning, development and management of the water resource projects, must be subjected to careful social scrutiny, based on well-developed mechanisms of accountability and regulations' (Goswami 2007).[1] This suggestion of the NWAC, however, was rejected by the Union Government without assigning any good reason.

While these observations in the media were informed by a broader level political economy perspective, the public consultations led by the popular peasant organisation KMSS not only did highlight the perils of the neoliberal economy but also articulated the specific concerns of the local communities including peasants and fisher folks. The major points of criticism against and comments on the draft which emerged from these public consultations and the discourses in the media may be summarised as the following:

- The underlying ideology of the policy to transform water, a traditionally enjoyed free natural resource, into a commodity cannot be accepted as it is anti-people.
- The policy encourages construction of large dams for production of hydropower and other purposes but it is completely silent on rehabilitation policy.
- The policy provides for flood control measures through mainly embankment which is not desirable. Embankments must be replaced by people- and environment-friendly methods.
- The policy encourages river linking schemes which will not only have a destructive impact on the ecology of the region but also on the livelihood concerns of the people in the valleys of these rivers.
- It is vague on restoring of natural water bodies like *beels*, ponds, etc.
- The policy is silent on women and their rights and role in water management.
- The policy will have a disastrous effect on the rural peasant society of Assam.

The civil society expressed its displeasure with the DSWP with public protests. They rejected the DSWP and demanded a new people-oriented policy through a democratic process by involving consultations with all sections of society including the most marginalised like the farmers, fisher folk, etc. In the meantime, civil society groups prepared an alternative draft of the SWP forcing the Government of Assam to come out with a revised policy by incorporating a declaration that 'No levy of any kind will be charged by the government or any agency for use of natural water by the people'. The other points in the draft, however, remained unchanged triggering more protest.

Negotiating the water policy

As the public protests persisted, the Working Committee on DSWP met a few of the civil society representatives and invited them to take part in the review meeting of the final DSWP. Before the review meeting, the civil society groups met twice in October 2008 in Guwahati, to discuss the points raised by civil society again in a larger forum. Most of the participants in these meetings launched a scathing attack on the DSWP and proposed an alternative water policy. These two consultations reiterated the earlier criticisms against the DSWP. The participants emphasised on giving primacy to community rights over using local water bodies such as rivers, wetlands, *beels*, tanks, etc. in the water policy. They also asserted that it was the government's responsibility to provide safe drinking water for free to each and every household. This responsibility was not to be handed over to any private party. Following the 27 October meeting, a series of negotiations took place between the civil society representatives and the government-constituted Task Force on the SWP from 30 December 2008 to 28 March 2009. It ensured the inclusion of one important point in the DSWP which highlighted the position of the community as the primary repository of rights to water. However, most of the critical issues still remain unresolved.

In this background, on 25 May 2009, the KMSS sent an ultimatum to the Task Force reiterating its demand on the following:

- The government must ensure a daily supply of 135 litres of free domestic and drinking water to each individual.
- The government must supply free irrigation facility to the agricultural land of the marginal farmers (up to 12 *bighas* per family).
- Obtaining the approval of the government and the people of Assam must be made mandatory in case of construction of big hydel projects in a different State or country on a river which flows through Assam.
- Obtaining of the prior approval of the local community (down to the level of Gram Sabhas) must be made mandatory in every stage of planning, implementation and management of schemes relating to water resources.
- Special schemes should be taken up for the protection of the degrading water bodies such as rivers, *beels*, ponds, marshes, etc.

In an article strategically published on the eve of a critical public meeting held on 4 June 2009, Akhil Gogoi, the General Secretary of KMSS, highlighted the fact that the DSWP was based on the National Water Policy 2002 even as according to the constitution 'water' is a State subject. He criticised the policy as a ploy to take away the resources of the State to serve the interest of big capital under the neoliberal development paradigm of the Indian State (Gogoi 2009).

The main points which emerged in the 4 June 2009 meeting were included in a memorandum which was submitted to the Chairman of the Task Force. These points included the following:

- Access to water supplied by the public authority should be divided into two categories: (1) water for drinking purpose; (2) water for commercial and industrial use.

- The supply of free drinking water is the responsibility of the State government up to 135 litres per person per day.
- Progressive water tariff will remain confined to the commercial use of water to the extent possible excluding the exempted categories.
- Cultivable land up to 12 *bighas* must not be subjected to any tariff for irrigation facilities.
- The first right to groundwater will be of the concerned community level organisations as per the 73rd Amendment of the Constitution of India and not individuals on land-ownership basis.
- The matters related to control over water resources in neighbouring States and on the upstream of rivers that flow through Assam needs to be carried out with due consultation and permission from the State of Assam, which may be affected by such initiatives.
- Special schemes should be undertaken to restore and preserve natural resources like ponds, *beels*, tanks, etc., by the State government.
- The provisions pertaining to river linking project should be removed from the policy.

Based on this memorandum, some points were included in the policy in the last meeting of the Task Force held on 17 June 2009. This became the final DSWP which was then submitted to the Government of Assam. However, an array of issues and concerns regarding the provisions of the draft remained unaddressed. For example, the provisions such as introduction of water meter on all consumers, progressive power tariff, etc. were not removed from the final draft. The latter also left the 'exempted categories' who would not be levied any charge for using water undefined. Besides these, the three issues which remained the focus of much of the conflict surrounding the DSWP pertaining to the river linking projects, new major hydropower projects and regional cooperation for construction of mega dam in the neighbouring States on interstate rivers remained unaddressed and unchanged in the policy.

Reacting to the final draft of the SWP, Hiren Gohain, noted public intellectual, held that the DSWP appeared to be more inclined to serve the interest of the big capital at the expense of the masses of peasantry and other marginal groups. He pointed out that much of the responsibility for the wastage of water in our country lies with the big industries. If they are entrusted with the responsibility of developing and managing the water resources of our society, the common masses would come under relentless repression with all kinds of taxation in the interest of big capital. He argued that the provision of public–private partnership is nothing but an instrument to surreptitiously hand over the crucial community resources to big capital. Although the draft repeatedly speaks of people's participation, there is every possibility that this would be reduced to a unilateral affair in view of the fact that various community bodies such as *gaon* (village) panchayats, municipalities, etc. are often dominated by the representatives of the ruling political parties. The concept of the 'Water Users Association' mooted in the DSWP, Gohain asserted, is also bound to fail in the absence of a democratic organisational structure. He alleged that one of the most significant features of the DSWP is its tendency towards bureaucratising the implementation of its various provisions. The institution of a State Water Resources Board/Council with the Chief Minister of Assam as the chairman is a case in point. In such bodies, the voice of the people cannot be expected to be given due respect (Gohain 2009).

Conflicting crystallising positions

It is evident that many groups and individuals from different sections of society in Assam participated in the debates and protests surrounding the DSWP. While the peasant and village-based organisations like KMSS was in the forefront of the public protests, the concerns of these protests were not confined to the issues of agriculture alone. Some of these organisations articulated larger issues of Centre–State relations, dangers of neoliberal economic policies and so on. They also expressed concerns about the environmental threat posed by the commercialisation of water, river linking schemes and large hydel projects. The civil society organisations from the urban areas representing mainly the poor and lower middle class also strongly articulated their concerns regarding introduction of water meter, the provision of progressive tariffs on the use of water and so on. They also found support from a section of middle class intelligentsia including academics, lawyers, technocrats and senior journalists as well as many environmental NGOs. In brief, the identity of the protesters against the DSWP transcended various social and spatial boundaries and assumed the character of a popular movement.

However, the policy also found support among a section of people. This section evidently included industrialists, ruling party politicians, bureaucrats, technocrats and so on. Yet, only a handful of individuals mainly retired bureaucrats and technocrats from this group participated in the public consultations. These individuals were invariably at the receiving end of public ire in the tumultuous public consultations. Interestingly, the established opposition political parties of the State were not much active on the issue. Their response to the water policy oscillated from symbolic protest to indifference.

It is also important to note that although several organisations representing peasants and other marginal rural communities were involved in the protests against the DSWP, these protests took place mainly in the Guwahati city and not so much in the rural areas. The poor rural communities took part in good numbers in the protests held in the city who were brought there by various organisations representing their cause.

Present status

The final DSWP incorporated some people-friendly provisions under pressure from the civil society organisations. However, many other critical provisions in the draft remained unchanged despite popular demand. As mentioned above, the Department of Water Resources of the Government of Assam and the Task Force gave final shape to the draft of the SWP in late June 2009 and handed over the same to the Government of Assam. However, since then, the latter is maintaining silence on the document and the protest against the water policy *per se* has become dormant, although a popular protest movement against the proposed and under-construction hydel dams in Arunachal Pradesh reached its heights in the Brahmaputra Valley during recent times.

In the meanwhile, in early 2012, the Draft National Water Policy 2012 was released by the Union government which came under serious criticism by the civil society. Besides, the river linking project also received a new fillip around that time after the Supreme Court of India directed the Union government to expedite the implementation of this project despite stiff opposition to this from a significant section of the people and experts (Venkatesan 2012). In March 2012, the Assam government got the Assam Ground Water Control and Regulation Act 2012 enacted in the State Assembly with the objective of regulating and controlling the use of groundwater in the State, which is depleting in many parts of Assam, mainly because of over exploitation (Handique 2012). Subsequently, the new National Water Policy 2012 was adopted by the National Water Resource Council at its meeting held on 28 December 2012. It emphasised commercialisation of water, transfer of State's role as provider of water to private bodies, etc. Though many States opposed the new policy, Assam accepted it.

Since mid-2012, citizens in Guwahati are up in arms against certain steps of the Government of Assam relating to water supply in the Guwahati city. These include its decision to transfer the responsibility of water supply in the city from the Guwahati Municipal Corporation to a newly constituted Guwahati Jal Board under a Metropolitan Drinking Water Supply and Sewerage Board Act, 2009. The Jal Board, chaired by the Minister of Urban Development of the Government of Assam, is a bureaucratic body without any representation from public. The citizens are in vehement protest against these moves which they perceive to be clear indication of government's attempt at privatisation of water supply leading to a hike of the water tariff. Thus, it seems that though the government of Assam is silent on the SWP, it is either engaged in surreptitiously implementing various provisions of the policy in an incremental manner or waiting for the excuse of a national policy or Act which would make the SWP a *fait accompli* for the people of the State. In fact, personal and unofficial communication with some key officials of the State Water Resources Department reveal that the DSWP of Assam has now become null and void since it could not be officially adopted within a stipulated time.

Way ahead

There is no doubt that there is a genuine ground of mistrust and apprehension among the people about the draft water policy which cannot be forcefully brushed aside by any overriding law or regulation. The government seems to be engaged in an attempt to wear out the movements using its dilly dallying tactic. Although it repeats the importance of dialogue at every instance, it is often engaged in scuttling these dialogues as the consultations and dialogues on the DSWP in Assam have shown. On the other hand, the space for judicial resolution of the conflict also seems to be progressively shrinking with new Acts and Regulations on water-related issues being brought into force.

In such a situation, the most effective instrument for resolving the conflict seems to be to put pressure on the government through democratic popular struggle to legislate a water policy that addresses the genuine concerns of the people. Along with this, the scope for using the available legal options also should be examined. This is important in the context of the fact that the Government of Assam, as mentioned above, seems to be following a strategy to implement various provisions of the DSWP bit by bit as the example of the Guwahati Jal Board has shown. However, a dialogue is still the best way to resolve the conflict and the possibilities of such a dialogue must be explored. The media can play a crucial role in creating a space for such dialogue by bringing together the different stakeholders to the negotiation table besides playing the role of a watchdog.

Note

1 *Ibid.* (Goswami quotes here from an article by Mihir Shah, Member of the National Water Advisory Board, published in *The Hindu* on 7 June 2002.)

References

Dutta, I. 2007. *Jalanitir Khachara Khon Manahpoot Nohol* (The draft water policy cannot be approved of). *Dainik Janambhumi*. 29 September 2007.

Gogoi, A. 2009. *Rajyik Jalaniti: Tarun Gogoi Sarkarar Sehotia Ku-kirti* (The state water policy: The latest misdeed of the Tarun Gogoi government). *Asomiya Pratidin*. 4 June 2009.

Gohain, H. 2009. *Ganamukhi Jalanitir Prahasan* (The irony of the people-oriented water policy). *Asomiya Pratidin*. 27 July 2009.

Goswami, S. 2007. *Rajyik Jalanitir Khacharar Aarar Katha* (The facts behind the draft water policy) (in two parts). *Dainik Janambhumi*. 16 and 17 September 2007.

Handique, B. 2012. Dry spell sparks water shortage fear. *The Telegraph*. 17 March 2012.

MoWR. 2002. National water policy, 2002. *Ministry of Water Resources, Government of India*, April 2002, p. 9.

Sharma, D. 2007. *Etopa Pani Mor Ajanma Adhikar* (A drop of water is my Birth right). *Dainik Janambhumi*. 6 and 7 October 2007.

Venkatesan, J. 2012. Set up special panel on linking of rivers forthwith, supreme court tells centre. *The Hindu*. 27 February 2012.

Websites

http://sandrp.in/wtrsect/Assam_Water_Policy_Concerns_of_RVC_Assam_Aug2007.pdf (Accessed on 24 January 2014).

www.ielrc.org/water/doc_states.php (Accessed on 12 December 2013).

Authors' association with the conflict

The author followed the developments ever since the DSWP was circulated for public consultations. He was also present and participated in some such consultations besides regularly interacting with the leaders of different organisations who opposed the draft. The author also viewed the protests against the DSWP as a form of social movement, an area of his own research interest.

7 Water conflicts in Northeast India

The need for a multi-track mechanism

Nani Gopal Mahanta

While the management of water as a resource in India is democratic in spirit, the execution of the water policy is completely centralised, technocratic and market driven. Water management is dominated by bureaucrats, contractors, private engineering firms and global financial institutions such as the World Bank and the Asian Development Bank (ADB). This is notwithstanding the fact that, both constitutionally and institutionally, India has some of the most democratic provisions for managing water resources and the environment. However, in practice India continued with the command model of water governance inherited from the colonial period. In the era of globalisation, this model has become further more opaque and centralised. The pursuit of a neoliberal agenda in the 1990s and after has intensified a command structure dictated by the market, and a quest to catch up with the economies of Asia with a higher growth rate, led by China. In the process, what has been sacrificed is democratic participation, accountability, rule of law, community ownership of resources and federalist principles of governance.

In response to the developmental and cultural policies of the Indian nation-state, violent political movements have emerged in the Northeast region (Map 7.1) which is oriented towards bringing about either a partial or a total change in the existing relationship, values and norms. As politics is about 'who gets what, when and how', such policies have greatly annoyed the young generation which had little faith in the nonviolent mode of protest. From the very beginning, there was a strong feeling that New Delhi wouldn't listen to the voice of the periphery, and would not give the community an opportunity to exercise control over its resources.

For example, the United Liberation Front of Asom (ULFA) has said,

> In the economic sphere, India has been engaged in large-scale exploitation. Despite its rich resources, Assam remains one of the

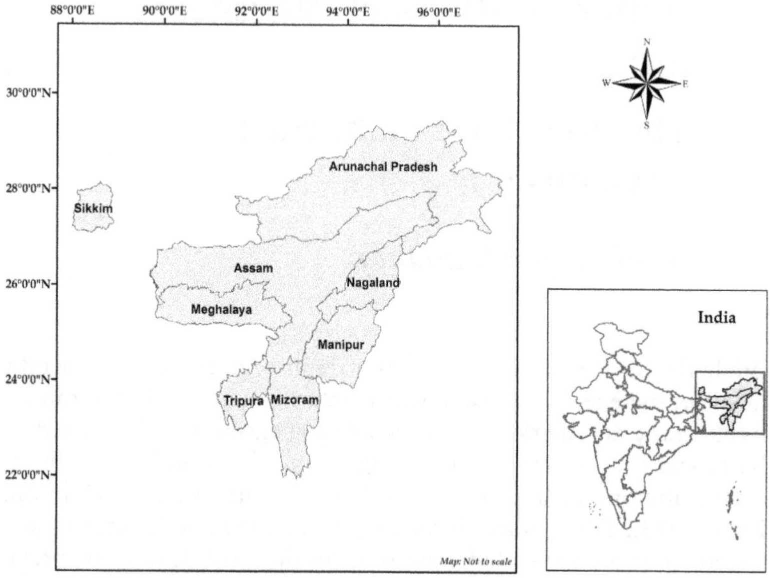

Map 7.1 The eight States of Northeast India

Source: SOPPECOM, Pune

> most backward states. Therefore, the question of a real threat to the national identity of the people of Assam under the colonial occupation and due to the exploitation by India has become the basic problem. As a whole, the problem has become a question of life and death for the people of Assam.[1]

Thus, a lack of control over the resources of the region and their plunder for the benefit of other metropolitan areas remain the central themes in the process of the crystallisation of Assam's identity.

The Northeast as a future powerhouse of India: the rationale

India's Northeast has been identified by New Delhi as the country's future 'powerhouse', and Arunachal Pradesh is slated to be the major contributor. In 2001, the country's Central Electricity Authority carried out a preliminary ranking of the hydroelectric potential of various Indian rivers. It identified 168 large projects in the Brahmaputra

Box 7.1 The story of oil and tea: who decides the fate of the resources in Assam

- Until 1977, Assam continued to produce half of the total crude oil produced in India. Assam also continued to produce half of the total natural gas produced till 1979. In spite of Assam's increasing crude oil production, the Government of India was not interested in establishing a new refinery in Assam.
- India decided much against the popular will to establish the refinery at Barauni in Bihar to process the crude oil produced in Assam. In order to transport the crude from Upper Assam oil fields to Barauni, a 1,400 km long pipeline which is one of the longest in the Third World was constructed during the early sixties.
- Russian experts were against such a move, and so was the 36th Lok Sabha committee.
- Assam, the highest producer of tea in India, accounts for 65 per cent of the total production. Nevertheless, States like West Bengal benefit more in terms of revenue, head offices, ancillary industries, auction centres, etc.
- The approach of the State towards the tea industry is still predominantly colonial, characterised by a lack of even marginal investments to develop the economic infrastructure of the region. Areas immediately surrounding the highly profitable tea estates continue to live in medieval darkness, while the managerial staff flaunts a lifestyle reminiscent of colonial days.

Basin alone, which collectively could generate more than 63,300 MW (Megawatt) of hydropower. Of these projects, as many as 87 were in Arunachal Pradesh. According to the October 2001 Central Electricity Authority (CEA) 'Preliminary Ranking Study' of the nationwide potential of hydroelectric schemes, it is the Brahmaputra River system which has the greatest potential.

The 168 schemes considered by the ranking study have a cumulative installed capacity of 63,328 MW, and 149 of these were given ranks A and B, indicating high viability. These schemes will be developed by agencies such as the National Hydro Power Corporation (NHPC), the North Eastern Electric Power Corporation (NEEPCO), the Brahmaputra Board and State Electricity Boards, and a major portion of this power will be evacuated to other parts of the country.

The rationale for the projects are:

- The projects will utilise the country's largest perennial water system to produce cheap, plentiful and renewable power for the entire country.
- At the local level, these installations will offer economic benefits through power export across the country.
- The projects will also provide employment opportunities.
- The projects will facilitate flood control.
- The projects will resolve conflict.

How big dams could lead to further conflicts in the region

There are five potential areas which if neglected could lead to a more volatile situation in the region:

1. Big dams and displacement
2. Cultural displacement and demographic change
3. Further militarisation of an already securitised zone
4. Marginalisation of the society and its aspirations
5. Human rights based development and indigenous issues

Big dams and displacement

Displacement is a continuous evil that accompanies any big dam project and history of the dam sites in Northeast bears testimony to this. Let's consider the case of the Pagladia dam which was planned in 1968 as a flood control scheme at a cost of Rs. 12.6 crore. The Public Investment Board of the Union Cabinet approved it at a cost of Rs. 526.62 crore in March 2000, and the Cabinet Committee on Economic Affairs approved it in November 2000 at a cost of Rs. 542.9 crore.

The purpose was to protect 40,000 hectares of land from floods and erosion, covering 190 villages, irrigating 54,160 hectares in 145 villages and generating 3 MW of electricity. The project was scheduled to be completed in 2007.

People who were to be affected by the dam were not even informed about it. They realised that a dam would be built only when the survey team came to their village. The Pagladia dam will submerge 34,000 acres of fertile and highly productive agricultural land.

The affected people were promised land in return, and a house has been built in the model village. When the people visited the resettlement site, they realised that the land was sandy and infertile. Besides,

Table 7.1 Likely displacement around some of the dams in the region

Dam	*State*	*Number of people to be displaced*
Gumpti	Tripura	60,000–70,000
Dumbur dam	Tripura	35,000–40,000
Loktak hydel	Manipur	20,000
Tipaimukh	Manipur	40,000
Pagladiya Dam	Assam	120,000
Siang, Dibang, etc.	Arunachal Pradesh	35,000–40,000

Note: This does not account for all 165 dams that are going to come up in the region.

Source: various newspapers in Northeast India

one of the two plots is already occupied by the East Pakistan refugees of 1947 and others who came later. Therefore, resettling the people displaced by the Pagladia dam on the same plot will only lead to major conflict.

The most important fact is that the Rs. 47.89 crore rehabilitation package is for 18,473 persons from 3,271 families, while the people claim that around 120,000 persons from 20,000 families in 38 revenue villages in the Tamulpur and Baganpara revenue circles will be uprooted, 90 per cent of them are tribal. Where will they go?

From 1968, the people facing displacement have been protesting against the dam and have formed an organisation called the Pagladia Bandh Prakalpar Khatigrasta Alekar Sangram Samittee. One of its leaders, Keshab Rajbangshi, reported that the protest is needed because the dam will uproot the already marginalised indigenous people both tribal and non-tribal living in the north bank of river Pagladia.

The extent of displacement can be gauged from Table 7.1, which states the number of people to be displaced by a few of the projects in the region.

Cultural displacement and demographic change

A fear of being swamped by outsiders is one of the most important causes for identity exclusiveness in Northeast India. In Tripura, Meghalaya and Assam, the fight against migrants remains the central theme shaping collective identity in the region.

Two States, Sikkim and Arunachal Pradesh, were relatively free from such movements against outsiders. However, the Teesta dam in Sikkim has already led to the influx of about 50,000 migrant workers from outside Sikkim.

It is feared that such migration will change the demography of the project area. The government has stated that certain precautions will be taken. The first preference will be given to the local people for all skilled, semi-skilled and unskilled work. Only when such workers are not available locally, will the developers be allowed to bring in workers from outside the State. All those who are brought in will have to be registered, and their work permit would be renewed every six months. Such labourers will have to leave the project sites as soon as their job is done.

Residential quarters and colonies shall not be allowed to be set up in the Dzongu area. This is a stipulation of the Ministry of Environment while issuing the environmental clearance.[2] However, there are few takers for these assurances.

Dawa Lepcha had been on a fast since 10 March 2008. In 2007, he fasted for 63 days. He and his friends are protesting against the dams on the Teesta River in Dzongu in north Sikkim, the home of the Lepchas, Sikkim's earliest inhabitants. These young men were in hospital, starving to make sure their tribe survives.

'The entire Teesta river is being tunnelled. The main river of Sikkim is disappearing underground. Is this development?' asks Dawa Lepcha.[3] 'Sikkim is a very small State, but very rich in biodiversity. If the government is allowed to go ahead with the hydel projects, they will ravage, plunder and destroy everything.' The Lepchas are the indigenous people of Sikkim, but they constitute less than 7 per cent of its population. They call themselves *rong-kup* (people of the ravine). Over the past two centuries, Nepali migrants have outnumbered the Lepchas in their homeland. The Lepchas are now a minority, a dying race. The Lepcha population is now 40,000, of which around 7,000 live in Dzongu, near the magnificent Kanchenjunga, the third highest mountain peak in the world.

Dzongu, the holy land of the Lepchas, was a protected area even when Sikkim was an independent kingdom (Sikkim merged with the Union of India in 1975). Any outsider, even a Lepcha living outside Dzongu, has to apply for a permit to enter it. Only Lepchas from Dzongu can own land here. This was done to protect the sacred land, the dying community and its culture. 'By building seven dams in the Lepcha-protected area, and allowing such a large influx of migrant labour, the government is violating its own laws. There are only 7,000 Lepchas in Dzongu. With just one project, we will be outnumbered. Our culture is under threat,' says Dawa Lepcha.

Similar is the scenario in Arunachal Pradesh. Mite Lingi, General-Secretary of the Idu Cultural and Literary Society (ICLS), the apex body of the Idu Mishmi community in the Dibang Valley, says that the State government kept local communities 'in the dark' during preliminary planning. Following the early May revelation of the State government's plans, Lingi's major concern was the number of external labourers who would have to be brought in for the new projects over a long period of time, that is, cumulatively at least 20 years. The population of the Idu Mishmi community is just about 12,000, explains Lingi. 'Thirteen projects will bring in at least 100,000 labourers. These are long-gestation projects, and we are seriously concerned about demographic changes in the Dibang Valley.'

Dam, militancy and militarisation of an already militarised zone

Various civil society groups in the region believe that the militarisation of dam areas will further accentuate militancy in the region.

- In the month of March 2008, two public hearings for the proposed Tipaimukh Multipurpose Hydroelectric Project, one at the Tipaimukh dam site, Churachandpur District on 31 March 2008, and another at the Keimai village, Tamenglong District, Manipur on 26 March 2008, were organised by the Manipur Pollution Control Board.
- An overall assessment of the five public hearings from 2004 till April 2008 indicates that the affected people of Manipur and Mizoram have all rejected the construction of the Tipaimukh dam project and that the project proponents pursued the project vigorously with little care for the rights of the affected peoples. Consequently, the affected peoples are unanimous and consolidating their stances to protect their land and resources and to defend their sources of survival.

There are attempts to divide civil society between those who support the dam and those who oppose it. The manipulation of project proponents in public hearings is crystal clear. The project authorities' stage managed back door affirmative representations of 25 villagers at the Tamenglong public hearing on 22 November 2006, against the stance of the people of Tamenglong District opposing the project. Affected villagers were denied participation in the secretive meetings. Only some handpicked villagers were entertained during the two hearings. Such methods constitute an attempt to drive a wedge among the people of Manipur towards creating disunity, misunderstanding and conflict.

Marginalisation of the society, its needs and aspirations

With the initiation of developmental projects in Manipur, the failure of the State and project authorities to recognise the inherent rights of the indigenous peoples over their land and resources, and their mandatory right to be consulted, poses a serious challenge. It is imperative that the State and project authorities seek the approval of the affected people before carrying out any initiatives detrimental to their land, rivers, wetlands and other resources.

- The project proponents have failed to recognise that the livelihood and survival, in both physical and spiritual terms, of the indigenous peoples of Manipur, revolves around the sustainable use of and dependence on their natural resources, and that the severance of this interrelationship will only lead to impoverishment and perennial hardship. Para-military forces are deployed in large numbers to protect dam construction.
- People are increasingly sidelined, marginalised and targeted, often with brute force and the application of the rigid state apparatus, while justice is denied to them. In a heavily militarised society like Manipur, this will only lead to rigidification of the practices of impunity and conscription of civilian space.
- Instead of promoting human rights based development, the government is relying on the military for security, as military personnel are deployed in large numbers in armed conflict regions like Manipur with the pretext of combating insurgents.

Human rights-based development and indigenous issues

In these areas, the collective rights of the indigenous people are denied, because they are too small a group in the eyes of policy makers. Indeed, there are legal institutional agreements and covenants that have recognised and built upon collective rights. The Declaration on the Right to Development itself has recognised the collective right of peoples in its Article 1, which states that every human being and all peoples are entitled to the human right to development, and also to the right to self-determination, exercising 'their inalienable right to full sovereignty over all their natural wealth and resources'. The Declaration on the Right to Development states categorically in Article 2 that 'the human person is the central subject of development and should be the active participant and beneficiary of the right to development'.[4]

The Third World proponents of the right to development must also take serious note of the collective rights of a nation or a State, as an implication of the human rights approach to development. The exercise of those rights must lead to the realisation of the right of all individuals to development.

Adhering to the text of the Declaration would imply that:

(a) the effective participation of all individuals in decision making and the execution of the process of development, which would necessarily require (1) transparency and accountability of all activities, (2) equality of access to resources, and (3) equity in the sharing of benefits. These are essential elements of the process of development which make the right to that process a human right;
(b) now it must be clear that economic growth and development of a State or a nation does not automatically lead to this process of development. In fact, if very specific policies are not taken to realise such development, the economic growth of a State often tends to increase the concentration of income and wealth, making the rich richer, and though not always, the poor poorer.

The way forward

People in the region are highly dissatisfied with the way environmental clearances have been provided to some big dams. As we have shown above, public hearings have been largely orchestrated to suit the purpose of large dams. A technocratic approach has gained preponderance over one that prioritises people's welfare. This is evident from the recently concluded Assam Assembly debate in the month of November 2010, which was exclusively convened to discuss the large dams issue in the region. The ruling party members are already determined to go ahead with the Lower Subansiri Hydroelectric Project, and had justified mega dam projects in the region. The Expert Committee report, which had expressed many reservations about the proposed dam, was practically nullified by the appointment of another multidisciplinary expert committee to look into the technical aspects and the downstream impact. The people of the region are highly sceptical and apprehensive about the construction of 168 large dams, a majority of which are reportedly being built by private sector players.

Such surreptitiousness has given rise to anti-national feelings. A denial of water rights to people is repugnant to the statutory arrangement of the country. For example, the National Water Policy 2002 advocates a participatory approach for the management of water resources. Article 21 of the Indian Constitution is about the right to life. This has to be read with the provisions of the Directive Principles of State Policy. Article 37 recognises the principle of equal access to material resources of the community. These principles are considered to be fundamental in the governance of the country. Article 39(b) states that 'the State shall, in particular, direct its policy towards securing that the ownership and control of the material resources of the community are so distributed as best to sub serve the common good'. Article 51-A (g) assigns a fundamental duty to every citizen of India to 'protect and improve the natural environment including forests, lakes, rivers, wildlife and to have compassion for living creatures'.

Developing multi-track institutional arrangements

From a conflict resolution perspective, for resolving water conflicts, we believe in a win-win situation where opinions of all stakeholders are accommodated while pursuing the development agenda. It will be naive to make the people solely responsible for the entire issue of electricity generation and alternatives to dam construction. The technical aspect of the issue cannot be trivialised. At the same time, it will be equally naive to expect the State to completely abandon developmental or mega dam projects, as we are already caught in the 'development trap' to catch up with global economies. As Ramaswamy R. Iyer has said, 'One could be in total agreement that people's participation is desirable, but remain unconvinced that it is a guarantee of wisdom or that it will necessarily lead to a change in the role of technological inputs in the process of development.'[5]

Four issues are important here:

1. An acceptance of democratic principles while pursuing the development agenda as enunciated by the constitution and other statutory bodies of the country. All the actors including the state and civil society bodies must believe in these principles. Many civil society bodies too have a tendency to jettison the state as one of the important players. Such an approach could be self-defeating.
2. Developing an institutional mechanism where the EIA reports, public hearings and other concerns of both the builder and stakeholders are taken into consideration. Here, we believe that the District Council and other Panchayat Raj Institutions (PRIs) could be the primary units where these issues could be discussed and debated.
3. The World Commission on Dam's (WCD) recommendations about the construction of large dams could perhaps be the best principles that need to be followed by the state and civil society groups.

4 Recognition of the right to water and water rights as fundamental rights.
5 Constitution of a Northeast Commission on Water and Big Dams (NECWBD) as a regulatory and adjudicatory body. A representative body of the North East Water Resources Authority (NEWRA) has already been proposed which is opposed by States like Arunachal Pradesh.

Various bodies have to be statutorily involved for an increased participation of people in big dam projects. The District Council or Zilla Parishad is the highest body in the decentralised Panchayati Raj administration in India. The Panchayati Raj Act (extension to scheduled areas), 1996, provided comprehensive powers for the traditional management practices of community resources, and for safeguarding and preserving community resources, which invariably include water. Any decisions concerning dams should be discussed in the District Council with the participation of all stakeholders. While public hearings can be conducted at the panchayat level with the supervision of the Collector and district administration, the environmental impact assessment (EIA) reports and other issues pertaining to resettlement and rehabilitation (R&R), protection of biodiversity, etc. can be discussed and decided at the Collector level. Such granting of powers would greatly resolve the dichotomy between the sovereign rights of peoples and those of the State. For the smooth functioning of the District Council, the unit will be assisted by the district administration, State pollution control board and other specialised bodies. Provisions must be made for the adequate representation of civil society groups in the special sessions of the District Council. Moreover, the District Council must be taken into confidence while conceptualising, planning, monitoring and executing any dams in the area.

Statutory clearances under the Forest Conservation Act, 1980, and Environment Protection Act, 1986, shall have to be discussed here. Land acquisition details must be discussed in the Gram Sabha, which is the primary unit of Panchayati Raj institutions (PRIs). If required, these bodies should consult and call for the services of any technical personnel from the State or outside the State. Ramaswamy R. Iyer calls this process a 'statutory clearance' for big projects.[6] In this manner, the PRI will be able to provide a holistic assessment of projects accommodating the views of both the state and society.

The WCD has developed certain strategic priorities in decision making based on recognising the rights of, and assessing the risks to all stakeholders. Strategic priority 1 focuses on public acceptance of key decisions. Acceptance emerges from recognising rights, addressing rights and risks, and safeguarding the entitlements of all groups of affected people.

Strategic priority 3 says that the management and operation practices must adapt continuously to changing circumstances over the projects' life, and must address outstanding social issues. It implies that all outstanding social issues associated with existing large dams must be identified and assessed. Processes and mechanisms must be developed for the affected people and communities.

Strategic priority 5 stresses the recognition of entitlements and the sharing of benefits. Joint negotiations with adversely affected people will result in mutually agreed and legally enforceable mitigation and development provisions.

Strategic priority 6 envisages that the government, developers, regulators and operators meet all commitments for planning, implementation and operation of dams. Compliance with all applicable regulations, criteria and guidelines and project specific negotiated settlements must be secured at all critical stages of project planning and implementation.

All these valid suggestions of the WCD can be implemented at the PRI level for Northeast India. However, what will then happen to the trans-border State mega dam projects such as the Subansiri which criss-crosses Assam and Arunachal Pradesh? In such cases, joint sessions of PRIs or such bodies constituted for the purpose could be held. Such issues that involve more than one State could be referred to the NECWBD as a regulatory and adjudicatory body. This body could be an autonomous commission comprising of elected representatives, NGOs, civil society bodies and technocrats.

Finally, the recognition of the Right to Water and Water Rights as fundamental rights by the Constitution of India would perhaps go a long way in the democratisation of the water sector in India. Democratisation itself is a process of conflict resolution. The policy makers of India frequently make the mistake of winning the hearts and minds of the indigenous people (generally referred to as tribal) through the magic stick of development. However, the main issues in this context are those of the centrality of habitat, and the control of indigenous people over land, forest and water. If they are forcibly taken away or given to crony capitalists, it would certainly invite more extremism and other forms of violence.

These are some of the institutional mechanisms that could be explored for the resolution of water conflicts between the nation and the people. A conflict resolution approach does not believe in the complete justification of one stand or the negation of the other. The big dam controversy in Northeast India is essentially trapped between two extremes: at one end are the state and technocrats who justify the need for big dams, and at the other, the extreme grassroots movements, civil society bodies and affected communities, who reject big dams and want them replaced by community oriented projects.

Notes

1 www.geocities.com/CapitolHill/Congress/7434.
2 For details, see *The Sikkim Times*, Government of Sikkim's statement of Hydel policy, 13 July 2007.
3 For his comments, see Teesta's Tears, in *Frontline*, 25(12), 7–20 June 2008.
4 Sengupta, A. 2006. The Right to Development in Human Rights in the World Community: Issues and Action. In: Claude, R. P. and B. H. Weston (eds.). *Human Rights in the World Community: Issues and Action*. Philadelphia: University of Pennsylvania Press.
5 Iyer, R. R. 1990. Large Dam Debate: A Response to an Intervention, *Economic and Political Weekly*, 25, 24 March.
6 *Ibid*.

8 Whose river is it, anyway?

The political economy of hydropower in the Eastern Himalayas[1]

Sanjib Baruah

There has been increasing talk of Maoists 'infiltrating' the protests against large dams in Assam. Chief Minister Tarun Gogoi cites 'Intelligence inputs and Home Ministry reports' to support his claim.[2] An internal memo of the Ministry of Home Affairs (MHA) warns of a 'new "Red Terror Corridor" along the Assam-Arunachal border'[3] – a geographical belt that includes the construction site of the controversial Lower Subansiri hydropower project. The area has been the focus of protests against mega dams on the rivers of the Eastern Himalayas. The 2,000 megawatts Lower Subansiri plant is one of the many hydropower projects planned in the region, and the first of the big ones to move into the construction phase. In 2011, a civil disobedience campaign began blocking trucks carrying building material to the dam site, which brought construction work to a standstill.

India's political and security establishment has been haunted by the spectre of Maoism for a while. But Northeast India has not been part of what is often referred to as the 'red corridor' – along which the so-called Maoist strongholds are supposedly located. The region has had its share of political militants. But they are motivated mostly by aspirations of group recognition and political autonomy, and they have been on the defensive lately. To what extent a Maoist movement is taking shape in the region is beyond the scope of this chapter. This chapter's focus is on the political economy of hydropower development in the region with particular emphasis on the Lower Subansiri project and the controversies surrounding it. The term Maoist, deployed by politicians and security officials in this context, I would suggest, functions as a counter-action frame: a strategic form of 'symbolic characterisation employed by oppositions against social movements' (Isaac 2008: 388; Haydu 1999). It carries the odour of un-patriotism, even treason, thanks to Maoism's Chinese provenance and India's strained relations with China. Thus when Chief Minister

Gogoi spoke of Maoists 'infiltrating' the anti-dam protests, he said that 'their acts stood to benefit China'.[4] 'They (protestors) do not wish Assam to develop. Other parties, including Maoists, are involved in the agitation,' said Gogoi in another context.[5] Indeed the suggestion that protesters are enemies of (national) development is an important sub-text of the Maoist counter-action frame. The debate on Lower Subansiri seems to have entered a new phase: security and intelligence agencies are getting ready to use coercive methods to quell the protests if necessary. In May 2012, security forces arrested many protesters and forcibly removed barricades and road blocks blocking the transportation of construction material to the dam site.

Legitimacy deficit

This turn of events underscores the serious legitimacy deficit of India's ambitious hydropower development plans on the rivers of the Eastern Himalayas. The great unevenness in the distribution of potential gains and losses – and of vulnerability to risks – has become rather obvious. There is talk of reliable and inexpensive energy attracting industries to Northeast India. But the hydropower that will be produced in the Lower Subansiri and other plants is meant almost entirely for use elsewhere, at least in the foreseeable future. Arunachal Pradesh, the 'host state', will be compensated handsomely with royalties from hydropower sales, and a small number of people in the immediate project area expected to be displaced in a physical sense will be compensated and rehabilitated. But official impact assessments give almost no attention to the serious threats to the livelihoods of the hundreds and thousands of people who depend on small-scale fishing and subsistence agriculture in the downstream areas of Assam and beyond. Equally controversial are serious geological hazards – seismic and hydrologic – specific to Northeast India that will add significantly to the burden imposed on the region. This lopsided distribution of costs, and of vulnerability, accounts for the serious legitimacy deficit in India's hydropower development policy in the Eastern Himalayas.

There is a fundamental difference between the hydropower projects of postmillennial India and the multipurpose river valley projects of an earlier period in India's post-colonial history. In the mid-twentieth century, large multipurpose river valley projects were taken up to develop a river basin region. They were driven by the spirit of decolonisation itself. The 'development of rivers', in Rohan D'Souza's words, had 'charged decolonizing nations with a new technological mission: the giant quest to transform fluvial powers into national assets – hydroelectricity, navigation, irrigation, and flood control' (D'Souza 2008). Globally speaking, that was a period when states were seeking to defend society against markets, and markets were regulated to promote the general welfare. That period has long passed. Hydropower development in post-millennial India is occurring under profoundly different conditions with 'state and markets . . . in collusion, jointly promoting the commodification of everyday life and the privatization of all things public' (Burawoy 2005: 154–155).[6]

The multipurpose river basin development projects of the past had used resources generated by hydropower to finance public goods such as flood mitigation, irrigation and navigation (Briscoe 2005: 34) and their focus was the development of a river basin region. By contrast, what is being designed and built these days are almost all single-purpose hydropower dams with power to be produced and sold for a profit by private as well as public sector companies.[7] 'In the Brahmaputra Basin', writes John Briscoe,[8] 'there are large benefits from multipurpose storage projects that are being foregone because power companies are licensed to develop "power-only" projects, which are typically run-of the river projects[9] with few flood control or navigation benefits' (Briscoe 2005: 34). The economics is quite simple: the fuel for hydropower production is moving water; and if society chooses to define property rights in a particular way, the owners of hydropower plants acquire the basic fuel almost free of cost. Given the 'free' source of fuel, unlike thermal power plants using coal, oil or natural gas, hydropower plants require huge initial investments, but once they are built, the operational costs are minimal. Hydropower is attractive to India also because of assured security of supply since increasingly the country has had to look for foreign sources of fossil fuel.

The power-only dams planned on the rivers of the Eastern Himalayas are fundamentally at odds not only with India's river-valley projects of the past, but also with recent authoritative pronouncements on visions of developing Northeast India's water resources. One such document is the strategy report entitled *Development and Growth in Northeast India: The Natural Resources, Water, and Environment Nexus* published as recently as June 2007 (World Bank 2007). The World Bank, several ministries and agencies of the Government of India and the governments of the Northeastern States collaborated on the report 'under the overall leadership of the Government of India, Ministry of Development of North Eastern Region' (World Bank 2007: x). The strategy report emphasised the need for 'a vision for the development of the Northeast as a whole' since 'the huge rivers, which are the lifeblood of the region, cut across several states'. The report articulates a vision of shared benefits for the region combining hydropower generation with goals such as reducing erosion of river banks – 'providing communities with assurance that investments in industries and infrastructure are sustainable' – and flood mitigation – 'saving millions of Assamese farmers from devastating effects on a recurring basis' (World Bank 2007: xvii–xviii).

In November 2004, Prime Minister Manmohan Singh had proposed the constitution of a 'cohesive, autonomous, self-contained entity called the Brahmaputra Valley Authority or the North East Water Resources Authority (NEWRA) to provide effective flood control, generate electricity, provide irrigation facilities, and develop infrastructure'.[10] The strategy report expressed strong support for this proposal. V.V.K. Rao, one of the water resource experts involved in the preparation of the strategy report, said that such a body 'could be the instrument for transforming the region' but for that to happen it would have to be 'cohesive, autonomous, self-contained' and have 'managerial and financial autonomy, [be] equipped with top-class manpower, and backed by Parliamentary sanction'.[11] However, the idea of such a statutory authority has gone nowhere ostensibly because of the opposition from the government of Arunachal Pradesh which 'gives no weight to flood control and navigation benefits (which would benefit the much larger populations in downstream Assam) and gives high weight to any submergence (which would displace people in Arunachal)' (Briscoe 2005: 34). If such an authority now comes into being, which is unlikely at the moment, it will have no more than a cosmetic function.

Clearly an autonomous statutory authority with a strategic vision of the development of the Brahmaputra River Basin as a whole should have been in place well before any major water resource development project was initiated. But given the perceived urgency of fast-tracking hydropower development, the Indian government has been aggressively pushing power-only dams after giving a rhetorical nod to the need for an authoritative body dedicated to the development of the river basin region as a whole. The results are predictable. As Don Blackmore, former Chief Executive of the Murray-Darling Basin Commission in Australia and another expert consultant involved in the strategy report, had warned, while an integrated approach could be 'truly transformational' for Northeast India, the existing institutional arrangements are 'simply too weak and fragmented to coherently manage complex river basins which cut across several different Indian States'.[12]

This short-sighted and potentially disastrous way of fast-tracking hydropower development reflects what Zygmunt Bauman has called the phenomenon of 'market pressures . . . replacing political legislation . . . as principal agenda setters'. It confirms Bauman's observation that 'a marked tendency in our times is the ongoing separation of power from politics' (Bauman 1999: 74–75).

Hydropower development and its discontents

Environmentalists favour hydropower because it is a low-carbon source of energy. Because water is replenished by the earth's hydrological cycle it is a renewable source of energy. But hydropower development, when it is large-scale, reeks of un-sustainability. Large hydropower dams on rivers of the Eastern Himalayas are sure to destroy the health of some of the world's most powerful rivers and their ecosystems. The adverse impact will be huge on the aquatic and terrestrial habitats of numerous plant and wildlife species, and it will have devastating consequences for the livelihood of communities that depend on them. The influential 2000 report of the World Commission of Dams (WCD) had concluded that while dams can bring 'substantial benefits', the record of dam-building is one of 'pervasive and systematic failure to assess the range of potential negative impacts' including the impact on 'downstream livelihoods'. The result is the 'impoverishment and suffering of millions, giving rise to growing opposition to dams by affected communities worldwide' (WCD 2000: xxx–xxxi).

In order to ensure that dams in future do not impose such heavy social costs, the WCD had proposed guidelines that break away from the notion of dam-building decisions being the exclusive domain of technocrats. The WCD advocated a participatory approach: treating the affected people as active negotiating partners and not as passive victims or beneficiaries (*ibid.*: 206–211). In addition, the report recommended a precautionary approach vis-à-vis decisions about dams: 'exercise caution when information is uncertain, unreliable, or inadequate and when the negative impacts of actions on the environment, human livelihoods, or health are potentially irreversible' (*ibid.*: 236–237).

The report was controversial. Activists engaged in the dam debates of the 1990s welcomed it, but dam construction and hydropower industries were unenthusiastic, as were many governments. Technocrats in India's Water Resources and Power Ministries and the public sector National Hydroelectric Power Corporation (NHPC) were among the report's most vocal critics. As Ramaswamy R. Iyer, himself a former secretary of that department, put it, the response of the Indian Ministry of Water Resources was 'comprehensively negative. . . . [T]heir comments are expressed in unusually strong language. This is not mere non-acceptance but total denunciation' (Iyer 2001: 2275).

Yet the WCD report was influential enough for many people to believe that mega dams had become a thing of the past. But in the twenty-first century, they have acquired a new lease of life for two reasons: (a) climate change has moved up on the global policy agenda, and the singular attention given to limiting carbon emissions has made hydropower more acceptable; and (b) the liberalisation and globalisation of capital and financial markets has opened up new ways of financing hydropower projects making the World Bank focused campaigns of the 1990s irrelevant. This was not unanticipated. For instance, Patrick McCully, author of the widely read book *Silenced Rivers*, had noted in the 2001 edition of his book that given the financial crunch that the dam industry was faced with:

> The great hope for the industry is that global warming will come to the rescue – that hydropower will be recognized as a 'climate-friendly' technology and receive carbon credits as part of the international emissions trading mechanisms under the Kyoto Protocol.
> (McCully 2001: xvii)

In the past one decade, McCully's comments have proved prescient.

There is by now a rich literature on the environmental impact of large dams on rivers, watersheds and aquatic ecosystems. To a significant extent it has informed the debate on the Lower Subansiri project. However, the discourse of environmentalism is of somewhat limited use when it comes to understanding – and mobilising public opinion on – the impact of large dams in 'developing' countries. The vocabulary of environmentalism may circulate globally. But it does not have the same political resonance among the general public in all parts of the world. While the laws of many countries, including India, require some version of environmental impact assessments (EIA) for major development projects, those legislations are not the result of domestic political pressures, but of a 'top-down international process'. Organisations such as the World Bank, the United Nations Environment Programme and the Organisation for Economic Co-operation and Development (OECD) had provided 'neat, standardized packages' of the EIA protocol to be incorporated into national legislations (Hironaka 2002). An EIA therefore means different things in different countries. In India, the scope of EIAs is extremely limited: it is a technical exercise with little scope for input from people affected by a project. Crucial parts of an EIA are done after a project is approved.

According to Ramaswamy R. Iyer, EIAs have been 'notoriously bad in this country and need to be substantially improved and distanced from the proposers, approvers and implementers of projects'.[13] Yet the EIA exercise provides some space for public criticism of projects. However, because of the priority that the EIA process appears to attach to the impact on endangered species rather than on humans, it has been possible for those pushing such projects to argue that only rich countries can afford environmental regulations, poor countries cannot. This has suited the interest of hydropower developers in Northeast India and their political supporters exceedingly well.

An older language of political economy can better capture the effects on people. I will argue below that the most important result of hydropower development on the rivers of the Eastern Himalayas is the 'enclosure of the hydro-commons' (Bakker 2003: 338), which will have a devastating impact on the lives of millions – especially the rural poor – that in a variety of ways depend on the water commons for their livelihood.

Great leap forward in hydropower and Northeast India

The failure to meet the demand for electricity has been the bane of India's economic growth story. Insufficient availability of power has substantially contributed to the slowing down of India's economic growth rate. It is estimated that India's power demand will rise by 350 per cent in two decades. That will require that the country triples its power generation capacity (World Bank 2007: 55–56). India is the world's seventh largest emitter of greenhouse gases, and the fifth largest emitter of carbon dioxide (CO_2) from fossil fuels. Coal-fired thermal plants are by far the most important source of electric power and it contributes the bulk of India's total CO_2 emissions. In global climate change forums, India is under growing pressure to reduce emissions. It has made serious commitments to reduce emissions per unit of gross domestic product (GDP). Therefore as India invests heavily in power generation, its energy strategy tries to balance between conflicting goals. During the 12th Five Year Plan period of 2012–17 India's target is to add an additional 100,000 megawatts of power generation capacity – an 'ecologically impossible' target, said former minister for Environment Jairam Ramesh.[14] 76.5 per cent of the targeted additional power is expected to be generated by fossil-fuel powered sources, and 20 per cent from hydropower.

Since the turn of the millennium, the intention of policy makers to fast-track hydropower development has been quite apparent. A vision document circulated by the Central Electricity Authority of India (CEA) in 2001 estimated the total cost of harnessing the remaining hydropower potential of the country by 2025–26, and provided preliminary ranking studies of about 400 hydropower dams with the total production capacity of about 107,000 MW. In order to interest potential developers, the CEA ranked the projects into five categories – A, B, C, D and E – in terms of feasibility by taking into account logistical as well as political constraints (in so far as they might involve interstate or international river disputes). In May 2003 the then Prime Minister Atal Bihari Vajpayee announced the 50,000 MW Initiative that included 'prefeasibility reports' on 162 new projects with an aggregate capacity of 47,930 MW (ADB 2007: 13). Those projects were to be completed by 2017, and were to be followed by another drive to add at least 67,000 MW of additional hydropower capacity in the subsequent 10-year period (Dharmadhikary 2008: 7). This projected pace of dam-building and the scale of India's hydropower development plans is unprecedented – nothing short of an attempt at a Great Leap Forward in hydropower generation.

A significant part of India's untapped hydropower potential is located in the rivers of the Eastern Himalayas. This has been known for a while. But the massive engineering projects necessary to tap into those reserves were not feasible till recently because of the region's poor communication infrastructure. According to CEA's estimates, Northeast India could generate as much as 58,971 megawatts of hydropower. Arunachal Pradesh alone has the potential of producing about 50,328 megawatts of hydropower – the highest in the country (Government of India 2004: 37). Government officials now project the State as India's future power house. A policy document of the State government proclaims that the State could be 'floating' on 'hydro-dollars', as oil-producing Middle Eastern countries presumably do on 'petro-dollars', if its hydropower potential is harnessed fully.[15] The expectation of the State's politicians that the State could make windfall gains from hydropower has facilitated New Delhi's drive to put hydropower development on the Eastern Himalayan rivers on a fast track.

As of October 2010, the government of Arunachal Pradesh had signed memoranda of understanding with developers on 132 hydropower projects with a total capacity of 40,140.5 MW; 120 of them are with private companies (Vagholikar and Das 2010). A news report of September 2011 puts the number of memoranda of understanding signed at 148 (*Down to Earth* 2011). According to one estimate, in a 10-year period, Arunachal Pradesh proposes to add hydropower capacity which 'is only a little less than the total hydropower capacity added in the whole country in 60 years of Independence' (Human Rights Law Network 2008: 3). While memoranda of understanding do not automatically translate into actual projects, the rush to draw up such agreements provides a glimpse into the process of fast-tracking hydropower development, in the absence of an autonomous river-valley authority with a strategic vision of what is good for the region. Potential developers pay significant sums of money to the State government of Arunachal Pradesh as advance payment 'before any public consultations, preparation of Detailed Project Reports and receipt of mandatory clearances' (Vagholikar and Das 2010). An activist of the Human Rights Law Network claims the following:

> Private players have made upfront payments to the government, which defeats the purpose of public hearings before beginning the construction of projects. . . . Upfront payment means the project becomes a fait accompli. The entire process under [the] Environment Protection Act of 1986 then becomes a farce since the government is in debt to the developer.[16]

Smelling the prospects of windfall profits, private companies have rushed in. Indeed companies 'unheard of in power and infrastructure sector' like Mountain Fall India, KSK Electricity Financing, Indiabull Real Estate, Raajratna Metal Industries have signed memoranda of agreement with the Arunachal government (Deka 2010: 4). In the case of the 1,750 MW Lower Demwe project on the Lohit River, Athena Energy Ventures has committed more than 15 per cent of free power to the State government as royalty[17] – significantly more than the 12 per cent that the public sector NHPC would have given.

In the rest of this chapter, I will focus on the two sets of issues that have been central to the controversy on mega dams in Assam over the past few years: dam safety and the impact on downstream livelihoods. But before that, let me first briefly introduce the Lower Subansiri Project and provide an overview of how the controversy has evolved.

The debate on lower Subansiri

One of the principal tributaries of the Brahmaputra, the Subansiri is Arunachal Pradesh's largest river system. After flowing through Tibet and descending down the Eastern Himalayas in Arunachal Pradesh the river flows into the Brahmaputra in the plains of Assam. The Lower Subansiri project will harness the hydro potential of the lower reaches of the river. The 116 m high concrete gravity dam under construction is located in 2.3 km upstream of Gerukamukh village in the Dhemaji District of Assam.

Protests against the Lower Subansiri project started as soon as the process of granting clearances under India's environment, forest and wildlife laws began in the early 2010s. Initially it was primarily environmental activists who had voiced opposition primarily on grounds that the dam site and submergence area includes a region known for its rich wildlife and biodiversity. They pointed out serious flaws in the EIA and other project documents submitted for obtaining clearances arguing that they underestimated the project's potential adverse impact on biodiversity. They raised procedural objections as well (Vagholikar and Ahmed 2003).

But opposition on other grounds also began to form during those early days. An activist was cited in an article published in 2003, saying that 'a small group of the downstream affected people have come together to form the Subansiri Bachao Committee [Save Subansiri Committee], but as of now, they have access to very little information'.[18] However, not all opposition came from those who objected to the dam in principle. There was also political mobilisation demanding that jobs, as well as subcontracts for the supply of construction material, go to locals.[19] The protests in Assam gained strength from around 2005 when construction began in earnest, and two influential State-wide organisations, the All Assam Students Union (AASU) and the Krishak Mukti Sangram Samiti (KMSS) – led by Akhil Gogoi, who had developed a nation-wide reputation as a right to information activist – got involved. Soon the State's main opposition party, Asom Gana Parishad (AGP) also joined in. The protests now were firmly focused on the issue of dam safety; and on the potential adverse impact on downstream communities, which had received little consideration in the approval process. The uneven distribution of the costs and vulnerability of the project also became an issue. 'Arunachal Pradesh is set to gain revenue from these projects', an AASU leader was quoted as saying, 'but Assam will be the victim if anything goes wrong'.[20]

As the anti-dam agitation gained strength, and the April 2011 elections to the State Assembly drew close, the Lower Subansiri dam emerged as a major political issue. Apart from the AGP, the Bharatiya Janata Party (BJP) also supported the protests. In July 2009, the Assam Legislative Assembly debated the issue and members of the ruling Congress party joined opposition Members of Legislative Assembly (MLAs) in expressing concern about the dam's potential adverse impact on downstream communities. A critical step in the controversy was the decision in December 2006 to constitute an expert group to study the downstream impact of the dam. It was the result of an agreement between the Assam State government, the NHPC and AASU, which by then was playing a major role in the protests. The expert group was constituted with eight academics – professors of Civil Engineering, Environmental Science, Geography, Geology, Life Sciences and Zoology at three of Assam's most prestigious academic institutions: Gauhati University, Dibrugarh University and the Indian Institute of Technology in Guwahati. The NHPC later indicated that it was a reluctant party to the agreement. It 'was constrained to award the study' to this group of academics, says an NHPC statement, and that the composition of the committee was 'suggested by AASU' (NHPC 2010a: 33).

The expert group submitted its report in June 2010. It confirmed that the fears expressed about the dam's safety and adverse impact on downstream communities were not unreasonable. The report recommended that the dam be redesigned: its height reduced and other changes made to increase the river flow and help flood moderation, but changes that would cut into its power producing capacity and profitability. The expert group's report galvanised the protests in Assam. By then the State's ruling Congress party was trying to present itself as sympathetic to the concerns of the protesters. As Chief Minister Tarun Gogoi said, 'We are not against dams', but if the Lower Subansiri leads to 'adverse environmental impact then we are not going to accept it'.[21] The State government, as one analyst put it, was 'in a bind. On the one hand, elections dictate it respect the popular sentiment. On the other hand, it cannot go against the Congress-led government at the Centre which wants dams' (*Down to Earth* 2010).

In the autumn of 2010, the political stakes were high enough for the then Environment minister Jairam Ramesh to come to Guwahati on 10 September and hold a 'public consultation' on the issue of large dams that extended for more than six hours. Activists and intellectuals associated with the protests interacted with the minister. In keeping with an assurance he gave to that audience, Ramesh wrote a letter to the Prime Minister conveying the sentiments expressed in that meeting. On the Lower Subansiri project, he said in that important letter, 'the dominant view in Assam appears to be that this project will have serious downstream impacts' and that they want the 'project to be scrapped completely'. However, he said he had 'made it absolutely clear to the audience that I am in no position to make any commitment on the existing Lower Subansiri which is under implementation. . . . All I promised was that I would convey the sentiments and concerns to the Prime Minister and the Union Power Minister'. But in the case of 'projects not yet started', he said that he made an assurance that 'we will carry out cumulative EIA studies as well as comprehensive biodiversity studies'. Ramesh conveyed his own views on the subject as follows:

> Personally, I believe that some of the concerns that were expressed cannot be dismissed lightly. They must be taken on board and every effort made to engage different sections of society in Assam particularly and in other North Eastern states as well. Right now the feeling in vocal sections of Assam's society particularly appears to be that 'mainland India' is exploiting the North-East hydel resources for its benefits, while the cost of this exploitation will be borne by the people of North-East.[22]

Those were heady days for the protests against mega dams. Since then the political environment has changed significantly. The Congress party scored an impressive victory in the Assam elections of April 2011, and Chief Minister Gogoi was reelected to power for a third term. In July 2011, Jairam Ramesh was shifted out of the Environment and Forests Ministry, while his rank was elevated to cabinet minister. There was little doubt that Ramesh being relatively pro-environmental, and the opposition of powerful industrialists and some of his cabinet colleagues, were behind the decision to change the leadership of the Environment and Forests Ministry.

Even before the elections, Assam government's ministers while maintaining a supportive stance towards the notion of modifying the dam's design were careful not to support the protesters' demand that work on the Lower Subansiri project be suspended. At the same time, they tried to slowly disassociate themselves from the recommendations of the expert group's report. Nor did they back the protesters' demand of a moratorium on approval of new hydropower project till concerns about the downstream impact of large dams are resolved satisfactorily. But by the autumn of 2011, Assam government ministers began saying that the protests were 'engineered' by 'Left-wing extremists' while at the same time drawing a distinction between 'civil society groups' involved in the protests and the 'extremists'.[23] The government's attitude towards protesters blocking transportation of construction material to the Lower Subansiri dam site also became confrontational. However, Chief Minister Gogoi has also been speaking of bringing in new experts, including 'international' ones, for advice on the matter. But meanwhile the Indian government has continued to give stamps of approval to other major hydropower projects.

Earthquakes, floods and the politics of risks

Geographers characterise Northeast India as a seismically active region where earthquakes impact the hydrologic characteristics and morphology of rivers and water bodies (Goswami 2008). In this region such changes occur not just in the vastness of geological time, but in the time-scale of ordinary humans. Historical sources are replete with references to earthquakes – including catastrophic ones. The following is an eyewitness account of the Great Assam earthquake on the night of 15 August 1950 that measured 8.7 on the Richter scale:

> [T]he main shock lasted five or six minutes. It was certainly of long duration and extreme violence, the motion being vertical, as though the crust of the earth were caving in. . . . The illusion of everything falling down an immeasurable shaft was, of course, heightened by rocks pouring down the mountain sides all round us with a fearful clatter. . . . [T]he vibration was so rapid as to suggest the roll of kettledrums. Dark as it was, we could see the ridges silhouetted against the paler sky, with their fuzzy outline of dancing trees. The noise was terrific, petrifying, and long continued as whole hillsides, studded with pine trees, slid into the valley.
>
> (Kingdon-Ward 1953: 172)

The account is by English botanist Francis Kingdon Ward who while on a 'plant-hunting expedition' found himself close to the earthquake's epicentre in Rima, Tibet. Some of those massive landslides – 'solid-looking hills . . . in the grip of a force which shook them as a terrier shakes a rat' as Kingdon-Ward (1952: 406) put it – blocked the downstream flow of the Dihing, Dibang and Subansiri rivers, tributaries of the Brahmaputra, and when the trapped water burst through in cascades a few days later, it caused catastrophic floods downstream. These are the same rivers on which major hydropower dams are being built or planned today. The earthquake and the floods of 1950 are deeply etched in the collective memory of the people of the Brahmaputra Valley. More than half a century later, villagers in the region remember the floods of 1950 as *Pahar Bhanga Pani* [hill-destroying floodwaters] and *Bolia Pani* [floodwaters driven by madness] (Das *et al.* 2009: 24).

The river regime of the Brahmaputra changed dramatically as a result of the 1950 earthquake. The phenomenal volume of landslide debris carried downstream raised the riverbed of the Brahmaputra. It went up initially by about 5 feet near the town of Dibrugarh, and as large volumes of debris continued to pour into the river, it rose by another 5 feet by 1955. Floods in the Brahmaputra Valley became more frequent and destructive ever since. Over the years, as rivers have changed course and riverbanks have eroded, thousands of acres of productive agricultural land, homes, valuable infrastructure and sometimes entire villages have been lost in Assam. The dykes and embankments built to provide flood protection have been no match to the Brahmaputra's fury. The region struggles with the consequences of the 1950 earthquake till this day. Floods have pushed thousands of people to new areas, and in recent years, to more and more vulnerable lands, to forest lands, protected areas, where they are often labelled encroachers. Many have been forced to give up agriculture and have moved to urban areas in search of alternative sources of livelihood.

It is not surprising that till recently, the focus of all policy thinking vis-à-vis rivers in Assam had focused exclusively on the issue of floods. In 1972 the Government of India had constituted the Brahmaputra Board with flood control as its primary mandate. Indeed a dam on the Subansiri was first envisaged as far back as 1955 under the auspices of what was then the Brahmaputra Flood Control Commission. It was envisioned as a multipurpose project with flood control and irrigation as primary goals. Indeed investigations on dams on rivers of the Brahmaputra River Basin including the Subansiri under the auspices of the Brahmaputra Flood Control Commission and the Brahmaputra Board were all for multipurpose projects: to provide flood control and irrigation, and also produce hydropower. But none of these plans got off the ground because of resource constraints. Now these plans have all been set aside in favour of single-purpose hydropower dams. In 2000, authority over the Lower Subansiri project was transferred from the Brahmaputra Board to the NHPC in order to make room for the power-only project currently under construction.

To the people of the Brahmaputra Valley, dams in the mountains and foothills that surround it, evoke a raw sense of danger and foreboding. Recent disastrous flash floods attributed to water released from two upstream hydropower plants – one in Arunachal Pradesh and the other in Bhutan – have accentuated the sense of vulnerability. There have been no public inquiries about the causes of these floods, nor any assurance that they won't occur again. Furthermore, the districts of Lakhimpur and Dhemaji have seen unprecedented levels of siltation, which locals attribute at least partly to the mining of the beds of the rivers for boulders, cobbles and pebbles for use in the Lower Subansiri dam, and another major construction project: the bridge over the river Brahmaputra at Bogibeel. Boulders moderate the currents in the rivers stopping sand from being washed away. The massive removal of boulders loosens the river sand and magnifies siltation. Sand-casting has overwhelmed many villages in the area in the past few years. Many see these developments as harbingers of a calamitous future.

'The selected site for the mega dam of the present dimension', concluded the expert group on the Lower Subansiri project 'was not appropriate in such a geologically and seismologically sensitive location'. From the 'geological, tectonic and seismological points of view', said the report, no mega hydropower project should be built in the 'Himalayan foothills, south of the MBT'.[24] The NHPC in its response asserted that 'the geological and seismological' aspects of the project were all 'thoroughly examined by specialists' during the project approval process. It pointed out that, top specialists of government agencies such as the Central Water Commission, the Geological Survey of India, Central Soil and Materials Research Station, and the Central Electricity Authority were involved in the process. The dam's location was cleared by the Indian government's 'highest authorities'. The seismic parameters of the design were approved by the National Committee on Seismic Design Parameters of River Valley Projects – 'the highest level committee' of the Indian government based on 'site-specific seismic design parameter studies' done by experts at the Department of Earthquake Engineering of the Indian Institute of Technology in Roorkee – 'one of the renowned institutes and has specialization in this field'. Furthermore, the NHPC pointed out that a number of large dams on the western Himalayas including the Bhakra dam on the Sutlej River in India and the Mangla dam on the Jhelum River in Pakistan have been built in a 'similar geological environment' as the Lower Subansiri dam and that they are 'functioning satisfactorily' (NHPC 2010b: 52–53).

Right from the beginning of the controversy, vernacular knowledge based on first-hand experience had presented a formidable challenge to expert knowledge. But the expert group's report added a new dimension to the controversy: there was now a contest over the authority of two rival bodies of expert knowledge. Whatever the prestige of the technical experts on India's top-level official agencies, their stamp of approval has failed to carry the day in the Brahmaputra Valley when it comes to issue of safety of dams in the Himalayas. By contrast, the geologists and civil engineers who teach in some of the region's best-known academic institutions and are members of the expert group, are seen as individuals who have more stakes in the safety of those dams than anyone living in Delhi or Roorkee. In local eyes, the report of the expert group has significant credibility. The comments of an Assamese hydroelectric engineer with a career in building and managing hydropower plants in the region behind him reflect this tension:

> The experts from IIT-Rourkee [which has India's premier Department of Earthquake Engineering] have not seen the earthquake-induced landslides of 1950 . . . when hundreds and thousands of trees floating downstream had nearly covered the Brahmaputra river. They have not seen that extraordinary spectacle. How can they say what a future disaster on the Subansiri might bring? If despite what we know from real life, we get seduced by what experts from Rourkee might say (about dam safety) even God Almighty may not be able to save the people of Assam.
>
> (Bhattacharyya 2010)

In January 2012, a paid advertisement by NHPC appeared in many newspapers in Assam under the heading 'Clearing the Myths by Presenting Facts on Subansiri Lower Hydro Power Project'. The advertisement juxtaposed a number of 'myths' against 'facts'. The arguments made in the expert group's report, with the words often lifted from the report itself, were presented as 'myths' – perhaps not the best way of showing open-mindedness regarding a contrary view. The first 'myth' listed was the expert group's argument that the site of the dam was inappropriate. The so-called 'fact' was as follows: 'The siting of the dam at the present location has been cleared by the designated authorities of Govt. of India, after satisfying themselves with the existing geological, seismological and tectonic setup at the site and adequacy of investigations.' Against the 'myth' that 'the dam may break' was another 'fact': 'NHPC has so far constructed 12 dams in different parts of the Himalayas, all of which are performing satisfactorily. In 36 years of NHPC history, there is no case of distress on any dam' (NHPC 2012). It is unlikely that any public relations strategist was behind this advertisement. Arrogantly asserting the superiority of expert knowledge produced by national experts over knowledge produced by regional experts, and labelling the latter as 'myths', isn't the best way of going about trying to bridge the trust gap on the issue of dam safety.

The debate has clearly turned into what Dutch social theorist Annemarie Mol calls ontological politics (Mol 1999). The impasse is a reminder of sociologist Ulrich Beck's argument that experts can only presume the 'cultural acceptance' of risks; they cannot produce that acceptance (Beck 1999: 58). The people of the Brahmaputra Valley seem to believe that they intuitively understand that there may be unintended risks to these large dams that are incalculable and uninsurable

(Beck 1999: 160n). To a significant degree, that feeling seems to be shared by experts and laymen alike. After all, 'acceptable risks are ultimately accepted risks', says Beck (1999: 58). And risk definition can sometimes boil down to a power game: 'some people have a greater capacity to define risks than others' (Beck 2006: 333).

The enclosure effect and downstream livelihoods

The building of large hydropower dams is bound to lead to the enclosure of the water commons (Bakker 2003), which in a region like Northeast India is bound to have a profound impact on the livelihood of riverine communities. In this section I will develop the argument with a few examples of the impact on downstream livelihoods. Rivers carry sediments. Indeed a river, it has been said, 'can be considered a body of flowing sediments as much as one of flowing water' (McCully 2001: 107). The blocking of sediment-borne nutrients is sure to negatively impact traditional flood-recession agriculture in the flood-plains, which is a major part of Assam's agriculture.

Second, dams obstruct fish passage, and it dramatically impacts the life cycle of many fish species. In Assam, fish is central to the people's diet and a major source of the caloric intake of poor people. It is hard to imagine fish surviving the power turbines of a hydropower dam. The changes in water temperatures, severe manipulation of water levels to meet the demands of power generation and the reduction of oxygen levels are not conducive to the migration and spawning habits of fish, and their growth and reproduction cycles. A recent World Bank study calls subsistence fisheries 'a vital but largely un-quantified economic activity and livelihood component of rural communities and particularly of the poor'. They 'provide vital local nutritious food and a safety net for many poor households'. It is not only an important source of nutrition for those who are fishermen in an occupational sense; fishing may be a component of the survival strategies of many rural households with diverse occupations. Yet 'because of their variety, dispersion and social complexity, small-scale fisheries are often poorly documented' (World Bank 2012). We know from comparative studies of dams, that they can have significant effects on fishing 'for many hundreds of kilometres downstream' (Adams 2000: 9). There are few places in the world where the cumulative negative impact of dams on flood recession agriculture in the floodplains and subsistence fisheries will be more devastating to the food security of rural communities – especially of the poor – than in Northeast India.

The expert group's report gives significant attention to these effects. The difference between the scope of this assessment and that of the EIA that was part of the official project-approval process is quite telling. The 'Study area' of the EIA study on the Lower Subansiri project was limited to 7 km upstream and 7 km downstream of the dam site. By contrast, the expert group, focusing on the downstream impact, surveyed the Subansiri River Basin downstream from the dam site quite comprehensively, dividing it into four sectors. It also considered the potential impact on the large numbers of wetlands that interact with the river in multiple ways including migration of various species of fish. The survey found that people in those zones depend on the river in many different ways depending on their distance from the river. Apart from getting water and fish from the river, they harvest fuel wood, sand and pebbles as well; and depend on it for transportation.

The diurnal and seasonal fluctuations in the water level of the river – the diurnal variations necessary to meet the variation of power demand during the day – says the report, 'will definitely affect the river ecosystem as well as the ecology of the connecting wetlands. Aquatic fauna and flora, and dolphin population of the Subansiri will be destroyed by the project with its existing design and operational parameters'. The fluctuations of the water level on downstream communities, the report concluded, will impact lives in the downstream areas in many other ways. For instance, it will dramatically affect the transportation infrastructure of country and motor boats that carry people as well as 'domesticated animals, crops, thatches, pottery articles, forest products'. The very low flow of the river during the day time in the dry season would restrict the movement of boats, while the sudden release of excess water from the reservoir when the rivers are in spate will make movement treacherous (LSEG Report 2010: Chapter x: 5–10).

The expert group's report made a number of suggestions for modifying the dam design in order to minimise those adverse impacts on downstream communities. They include provision of fish passes, efforts to improve the replication of natural stream flow by using hydrological data and the 'appropriate temperature and oxygenation of water released downstream' (LSEG Report 2010: Chapter x: 10). Such modification of the dam design will significantly add to the cost of the project and make the Lower Subansiri hydropower plant less profitable. Clearly the expert group has a very different kind of dam in mind – perhaps more consistent with the multipurpose river valley projects of the past – than what India's present hydropower development policy permits.

The NHPC's response to those parts of the report was dismissive. Indeed it took the report's portrait of the poor conditions of the villagers – with river bank erosion and floods regularly damaging crops and homes – as evidence that the traditional livelihoods are not worth protecting. The report, it argues, 'shows that even without the Project people are already facing hardships in the downstream area'. By contrast, it points at 'opportunities' presented by dam during the construction phase itself to those living closer to the dam site. 'It has been seen in general', says the statement, 'that the people residing in the vicinity of the project are benefited in terms of their socio-economic development, infrastructure development, and educational facilities, indirect employment, business opportunities etc' (NHPC 2010a: 44). That the nonviability of flood recession agriculture and the possible disappearance of fish from rivers and wetlands could only mean starvation for many of our poorer fellow citizens does not seem to occur to our experts at NHPC drafting such notes.

Apart from the changes in the river's flow regime, the rules of public access to the river will also change radically once a dam is built. There are already examples of this from hydropower plants now in operation. A hydropower plant on the river Ranganadi located at Yazali of Lower Subansiri District of Arunachal Pradesh and managed by the North Eastern Electric Power Corporation (NEEPCO) has been in operation for a decade. Kimin, a small village in Papum Pare District of Arunachal Pradesh, has been at the receiving end of this plant's water flow regulation regime. A number of times Ranganadi's waters have been released without warning, and crops have been destroyed and cattle washed away. One person has died. The people of Kimin are now afraid of getting too close to the river, and they don't let their children play on the riverbank. In July 2006, in response to complaints by Kimin's residents, and the controversies, the Yazali plant's management issued the following notification:

> [D]uring monsoon . . . the gates of the Ranganadi Diversion Dam may require opening from time to time. . . . All concerned authorities, village headmen are therefore requested to bring the same to the notice of all villagers . . . to refrain from going to the river and . . . also restrict pet animals from moving around the river/reservoir during the monsoon period. The corporation will not take any responsibility for any loss of life of human, pet animals etc.[25]

The use of the term 'pet animal' to refer to farm animals and livestock probably indicates the gulf that separates the urban middle class world of power plant managers from that of the rural poor. But that apart, the notification is an extraordinary document. Akhil Gogoi spells out its meaning this way. The power companies seem to believe that all through the rainy season when they release water, 'those of us living by the riverbank, should be prepared to run like monkeys to the nearest treetop to escape the rising waters' (Gogoi 2011: 192). The spirit of NHPC's response to the expert group's concerns in this matter is not unlike that of the notification quoted above. 'Regarding the fear for the sudden release of water downstream', says the NHPC, 'people will be duly informed as is practiced in every river valley project so that no loss is incurred in human lives and property' (NHPC 2010a: 44).

There is perhaps no better example of what post-colonial theorist Rob Nixon calls the 'resource law of inverse proximity' at work – 'the closer people live to the resources being "developed" the less likely they are to benefit from that "development"' (Nixon 2010: 75). The people living by a river, writes Nixon, 'may belong to the land but, within a Lockeian logic of private property, the land doesn't belong to them. Thus in terms of the right to remain (not to speak of the right to just compensation) they can readily be cast as uninhabitants, residual presences from a pre-capitalist era' (Nixon 2010: 74). In this discursive framework their loss of 'nonmarket access to the means of livelihood' (Araghi 2000: 146) becomes inevitable. This is the enclosure effect of hydropower dams. Like the enclosures of the eighteenth century, as historian E.P. Thompson had put it, what is at issue is 'alternative definitions of property rights' (Thompson 1975: 261). The results, as Karl Polanyi had said, 'have appropriately been called a revolution of the rich against the poor. . . . The lords and nobles . . . were literally robbing the poor of their share in the common' (Polanyi 1980: 35).

Conclusions

Prominent experts on water resources – some of them known for their staunch support of large dams – have expressed astonishment at the Government of India's political incapacity to put in place a statutory body that has the power, autonomy and professional competence to focus on the development of the Brahmaputra River Basin region as a whole. Among them is John Briscoe, whom critics call the World Bank's 'main large dam crusader' (Bosshard 2008). Indeed in the debate on the Lower Subansiri project, the NHPC itself cites his book *India's Water Economy: Bracing for a Turbulent Future* to make the argument that the expert group's study ignores studies showing the 'positive impacts of dams' (NHPC 2010a: 43). While Briscoe in that book indeed argues in favour of large water projects, when it comes to the Brahmaputra Basin, he laments the Indian Government's failure to find 'a formula for getting good multipurpose outcomes'. Briscoe contrasts this to

> many of the world's most successful river basin development programs – ranging from the legendary Tennessee Valley Authority of the 1930s to the present-day Yangtze Basin development project [that] have relied on hydropower to generate the resources necessary to fund 'public goods', such as navigation and flood control.

The Three Gorges Dam on the Yangtze, Briscoe points out, 'is operated as a flood control dam, at an opportunity cost of a massive $1.5 billion a year in foregone power revenues'. Unfortunately, despite there being 'a history of successful multipurpose projects in India', writes Briscoe, 'the Government of India now does not have an enabling framework which facilitates the same socially-optimal outcomes' (Briscoe 2005: 34).

What is occurring in the Brahmaputra Valley today is resistance by a riverine people refusing to accept the risk definition of powerful elites bent on pursuing a strategy of accumulation by dispossession (Harvey 2003) and trying to turn their rivers into free fuel for hydropower plants, in utter disregard of the impact on them. Yet it seems unlikely that the protests will stop the construction of the Lower Subansiri dam, or unravel India's plan to achieve a Great Leap Forward in hydropower. Far too powerful economic and political forces are arrayed against the protesters. Furthermore, the protests are unlikely to be immune to the sociological logic of all such protests – that public attention do not remain focused on any single issue for very long (Downs 1972).

Multiple factors will determine how the politics of hydropower in the Eastern Himalayas play out. The arguments about ecological impact and threats to downstream livelihoods will have to contend with competing arguments about the trade-offs necessary for promoting India's development. In today's India, the appeal of such arguments can hardly be underestimated even in Northeast India – especially among the urban middle classes that are relatively insulated from the impact of the enclosure of the water commons. They are willing to latch on to any hopeful sign of the possibility of leaving the conditions of 'economic backwardness' and 'insurgencies' behind. Moreover, it has become increasingly apparent that dam protesters in future will face more than intellectual arguments on the other side of the barricades. They will have to contend with the coercive apparatus of the state as well.

It is significant that a part of the WCD report which the Indian government had found most pernicious is the idea that 'local people settled on the river banks' are stakeholders, and its recommendation that decisions to build dams should be based on their 'free, prior and informed consent'. The critique of the report by Indian officials charged that the WCD's report had made 'a dam decision subject to veto power of the local people settled on the river banks'. The stakeholders, it argued, are not only the people living in the river valleys or those directly impacted by dams, 'people who are to benefit from a project are also to be considered as stakeholder'.[26] In this way of thinking the stakeholders in the Lower Subansiri and the other hydropower projects are not only the people living by the rivers in Arunachal Pradesh and Assam, but also potential consumers of hydropower – perhaps anyone connected to the national power grid including the power-hungry corporate houses of Delhi or Ghaziabad. The principal implication of the Great Leap Forward in hydropower for Northeast India seems quite clear: the region is entering the era of late capitalism in a familiar role – as supplier of a key natural resource to fuel the engines of economic dynamism elsewhere.

It has been said that if the proposed projects materialise, there will be a fundamental transformation of 'the landscape, ecology and economy' of the Himalayan region with 'far-reaching impacts all the way down to the river deltas' (Dharmadhikary 2008). We are in the very early stages of the process. Without regional and basin-scale cumulative impact assessments, it is impossible to talk precisely about the full ecological or socio-economic impact on the region. But there is particular reticence on the part of the Indian government to engage in such studies. In their absence, estimates of the impact can only be speculative. However, one thing seems quite certain, that if the plans are carried out as currently envisaged, the Great Leap Forward in hydropower will play as decisive a role in shaping Northeast India's future as tea, oil and coal did in the nineteenth century.

A recent assessment of the impact of hydro-development in the Mekong River Basin concluded that

> [I]nstead of an economic hydro-boom as anticipated by many, continued dam-building on the Mekong and its tributaries could result in a non-traditional security disaster characterized by severe food shortages, destruction of livelihoods, and large, irregular movements of people.

The study portrays a scenario of 'loss of livelihoods, decimation of fisheries, destruction of crops and general human insecurity', that 'not only threaten the economic growth of continental South East Asia, but risk ending the more stable political environment that the region has been witnessing in recent years' (Baker 2012: iv). One cannot rule out hydropower development on the rivers of the Eastern Himalayas leading to a similar 'nontraditional security disaster'.

Unfortunately, the people of the Brahmaputra Valley may not have any choice but to learn to live with the risks of mega dams. No one ever asked them if they are willing to exchange the hazards of natural disasters for the risks of modern industrialism. Yet what India's technocratic establishment has concluded are 'acceptable risks' will become inescapable parts of their lives – at least for those who can continue living in their existing villages and make a living. They will have to find the resources within them – culturally and psychologically – to deal with those risks. Their old ways of dealing with pre-industrial hazards won't do. 'No matter how large and devastating', as Ulrich Beck reminds us, they 'were "strokes of fate" raining down on mankind from "outside" and attributable to an "other" – gods, demons, or Nature'. What they will be faced with now are not hazards, but risks: the product of human decision-making. For Beck, risks are different from hazards because they 'presume industrial, that is, techno-economic decisions and considerations of utility' (Beck 1999: 50).

The people of the Brahmaputra Valley have been living with floods since time immemorial. They have been able to deal with flood hazards mostly by watching changes in the colours of the sky, and the conditions of rivers that they know intimately. Villagers told researchers studying flood adaptation in the region that when the sky turns dark with grey and black clouds, they expect rain within a couple of hours. When dark clouds appear on the horizon, and the distant hills become invisible, they know that it is raining in the hills, and that flood-waters would come. They have even learnt to predict them: in Majgaon they expect flood waters 3 to 4 hours after the rains in the hills, and 6 to 12 hours in Matmora: two areas that the research team had studied. The villagers keep a close watch on river conditions – the speed of the currents and wave size. Then they figure out the timing and the intensity of floods, and decide whether to stay in their homes, or prepare to move to higher grounds (Das *et al.* 2009: 24).

Once hydropower plants are up and running, this way of predicting floods will become hopelessly old-fashioned, and even dangerous. Flood-waters could rush to their homes and fields even under a clear blue sky. All it would take is for a power plant manager to follow standard operating procedure and decide to release a certain amount of water – though the 'people will be duly informed', as the NHPC tries to reassure us (NHPC 2010a: 44). Nor can anyone rule out another disaster in the hills bringing floodwaters driven by madness ['*Bolia Pani*'] to the plains. But this time no one will know for sure, whether it is the result of an act of God, or an act of Man. But were such a disaster to happen, one can be sure that some technocrat will remind people and I paraphrase Beck, that statements about risks are inherently ambiguous, nothing is established in terms of certainty: they are only calculations of probability. Nothing is ever ruled out. Risk assessments can only say that in the balance of probability certain things are less likely to happen. Thus those who respond to critics of the Lower Subansiri dam today by saying that the risks of a dam failure are near zero, can easily turn around and 'bemoan the stupidity of the public tomorrow, after the catastrophe has happened', for failing to understand the fuzzy calculus of risks (Beck 1994: 9).

Notes

1 This article was first published in the *Economic and Political Weekly* (Mumbai), 21 July 2012, Vol. 47, Issue No. 29, pp. 41–52. The author and the editors are grateful to the *Economic and Political Weekly* for permission to reprint the article.
2 'Maoists formed body to oppose dams: CM', *Assam Tribune*, 13 January 2012.
3 Cited in Sahni (2012).
4 'Maoists formed body to oppose dams: CM', *Assam Tribune*, 13 January 2012.
5 'Dams on Brahmaputra: Team Anna member on fast unto death', *Rediff News*, 19 May 2012 (www.rediff.com/news/report/dams-on-brahmaputra-team-anna-member-on-fast-unto-death/20120519.htm).
6 Michael Burawoy's argument (Burawoy 2010) is an extension of Karl Polanyi's classic argument on *The Great Transformation* (1980). Polanyi wrote on what Burawoy calls the first wave of marketization in the nineteenth-century Britain marked by the commodification of labour. It led to the counter-movement that consisted of reforms such as the recognition of trade unions, reduction of working hours and child labor laws – the First Great Transformation. The second wave of marketization began at the end of the nineteenth century with the expansion of imperialism. It was interrupted by the Second World War, but it proceeded with renewed vigor in the 1920s. The counter-movement led to the Second Great Transformation which included protections provided by a variety of regimes including fascism, Stalinism, the New Deal and Social Democracy. The Second Great Transformation also included decolonization, socialist planned economies and the developmental state in newly independent countries. Polanyi expected that the counter-movement against the second-wave marketization will be the end of market fundamentalism. However, contrary to

his expectations, says Burawoy, since the 1980s 'third-wave marketization is sweeping the earth, with the state no longer a bulwark to market expansion but its agent and partner. The state, either directly or indirectly through the market, takes the offensive against labor and social rights won in previous periods' (Burawoy 2005: 157).

7 Of the 147 hydropower projects for which the Arunachal Pradesh government has signed memoranda of agreement with developers only one is for a 'multipurpose' project – the 3,000 MW Dibang project is referred to as 'multipurpose' because it has a flood moderation component. The rest are all single-purpose hydropower projects.

8 Formerly of the World Bank and currently Professor of the Practice of Environmental Health and the Practice of Environmental Engineering at Harvard University.

9 The term 'run-of-the-river' projects can be misleading: it is used to refer to a variety of projects. Arguably, a hydro project is run-of-the-river only when 'inflow equals outflow on a real-time basis, i.e. if there is no storage or flow modification at all'. By contrast, the large hydro-projects in the Eastern Himalayas, while they are referred to as 'run-of-the-river', they 'involve large dams which divert the river waters through long tunnels, before the water is dropped back into the river at a downstream location after passing through a powerhouse. These projects are promoted as being "environmentally benign" as they involve smaller submergences and lesser regulation of water as compared to conventional storage dams' (Vagholikar and Das 2010).

10 Cited in V.V.K. Rao, 'Hydropower in the Northeast: Potential and Harnessing Analysis', Background Paper No. 6, 2006, p. 37. This paper by Rao was commissioned as input to the study that led to the strategy report (World Bank 2007).

11 *Ibid.*

12 Don Blackmore, 'Management Structures to Lead the Brahmaputra Basis into the 21st Century', Background Paper No. 10, 2005, pp. 2–3. Paper commissioned as input to the study that led to the strategy report (World Bank 2007).

13 Speech at the Convention on Dams, Guwahati, 27 February 2012.

14 'Jairam Raises Red Flag over Power Projections', *Financial Express*, 23 April 2011.

15 Government of Arunachal Pradesh, 'Hydro Electric Power Policy', p. 5 (www.arunachalhydro.org.in/pdf/State%20Mega%20Hydro.pdf).

16 Cited in *Down to Earth* (2011).

17 Minutes of the 24th Meeting of the Standing Committee of National Board for Wildlife, p. 18, 13 December 2011, New Delhi (http://moef.nic.in/downloads/public-information/mom-24-13.12.11.pdf).

18 Cited in Vagholikar and Ahmed (2003: 27).

19 'Overcoming the Odds', *Telegraph* (Guwahati), 27 September 2004.

20 Cited in *Down to Earth* (2010).

21 'Arunachal mega dam creates political storm in Assam', *Hindustan Times*, 27 October 2010.

22 Jairam Ramesh, Letter to the Prime Minister of India, Reprinted in Talukdar and Kalita (2010: 96–100).

23 'Assam Govt sees Maoist hand behind anti-dam protests', *Rediff News*, 7 October 2011.
24 LSEG Report (2010, Chapter x:10). The MBT or the Main Boundary Fault, a tectonic boundary between the Lesser Himalayas and sub-Himalayas, is located north of the Lower Subansiri dam site.
25 Cited in *Down to Earth* (2008).
26 M. Gopalakrishnan, Detailed Comments on the WCD Report, United Nations Environment Programme, Dams and Development Project. 2001 (www.unep.org/dams/documents/default.asp?documentid=469).

References

Aaranyak. 2009. Adjusting to Floods on the Brahmaputra Plains, Assam, India. In: *Local Responses to Too Much and Too Little Water in the Greater Himalayan Region*. International Centre for Integrated Mountain Development, Kathmandu, Nepal, pp. 43–49.

Adams, W. 2000. *Downstream Impacts of Dams*. Working Paper prepared for Thematic Review I.1 for the World Commission on Dams. http://oldwww.wii.gov.in/eianew/eia/dams%20and%20development/kbase/con-trib/soc195.pdf.

Araghi, F. 2000. 'The Great Global Enclosure of our Times: Peasants and the Agrarian Question at the End of the Twentieth Century'. In: Fred, M., John Bellamy Foster and Frederick H. Buttel (eds.). *Hungry for Profit: The Agribusiness Threat to Farmers, Food, and the Environment*. New York: Monthly Review Press, pp. 145–160.

Baker, C. G. 2012. *Dams, Power and Security in the Mekong: A Non-Traditional Security Assessment of Hydro-Development in the Mekong River Basin*. NTS-Asia Research Paper No. 8. Singapore: RSIS Centre for Non-Traditional Security Studies for NTS-Asia.

Bakker, K. 2003. Archipelagos and networks: Urbanization and water privatization in the South. *The Geographical Journal*, 169(4): 328–341.

Bauman, Z. 1999. *In Search of Politics*. Stanford: Stanford University Press.

Beck, U. 1994. 'The Reinvention of Politics: Towards a Theory of Reflexive Modernization'. In: Ulrich, B., A. Giddens and S. Lash (eds.). *Reflexive Modernization: Politics, Tradition and Aesthetics in the Modern Social Order*. Stanford: Stanford University Press, pp. 1–55.

Beck, U. 1999. *World Risk Society*. Cambridge: Polity Press.

Beck, U. 2006. Living in the world risk society. *Economy and Society*, 35(3): 329–345.

Bhattacharyya, T. C. 2010. *Namoni Xowanxiri Bandh Proxongo: NHPC-k Xomoi Diboloihe Bixesogyo Committee Gothon Kora Hoisil* [About the lower subansiri dam: Expert group was formed only to allow more time to NHPC]. *Axomiya Protidin* (Guwahati). 20 October, 2010.

Bosshard, P. 2008. 'Dam Crusader's Swan Song'. World Rivers Blog. www.internationalrivers.org/blogs/227/dam-crusader%E2%80%99s-swan-song.

Briscoe, J. 2005. *India's Water Economy: Bracing for a Turbulent Future.* Report No. 34750-IN, Agriculture and Rural Development Unit. Washington, DC: The World Bank.

Burawoy, M. 2005. Third-wave sociology and the end of pure science. *The American Sociologist*, 36: 152–165.

Burawoy, M. 2010. From Polanyi to Pollyanna: The falsepptimism of global labor studies. *Global Labour Journal*, 1(2): 301–313.

Das, P., D. Chutiya and N. Hazarika. 2009. *Adjusting to Floods on Brahmaputra Plains, Assam, India.* Kathmandu: ICIMOD, p. 57.

Deka, K. 2010. 'Dams: The larger picture'. In: Mrinal Talukdar and Kishor Kumar Kalita (eds.). *Big Dams and Assam.* Guwahati: Nanda Talukdar Foundation, pp. 2–9.

Dharmadhikary, S. 2008. *Mountains of Concrete: Dam Building in the Himalayas.* Berkeley, CA: International Rivers.

Downs, A. 1972. Up and Down with Ecology: The 'Issue-Attention Cycle'. *The Public Interest*, 28: 38–50.

Down to Earth. 2008. Project disaster. May 15. www.downtoearth.org.in/node/4537.

Down to Earth. 2011. MoU virus hits Arunachal Pradesh. 9 September. www.downtoearth.org.in/content/mou-virus-hits-arunachal-pradesh.

D'Souza, R. 2008. Framing India's hydraulic crisis: The politics of the modern large dam. *Monthly Review*, 60(3): 112–124.

Dutta A. P. 2010. Assam's dam crisis. *Down to Earth*, 15 October, New Delhi. http://www.downtoearth.org.in/news/assams-dam-crisis-1978 (Accessed on 1 May 2017).

Gogoi, A. 2011. *Morubhumi Ahe Lahe Lahe* [The Desert Comes Slowly]. Guwahati: Akhor Prokax.

Goswami, D. C. 2008. Managing the wealth and woes of the River Brahmaputra. *Ishani* (Guwahati), 2(4). www.indianfolklore.org/journals/index.php/Ish/article/view/449/514.

Government of India (GoI). 2004. *Mission 2012: Power for All, Powering India's Growth.* Ministry of Power, Annual Report, 2003–2004. New Delhi. www.powermin.nic.in/reports/pdf/ar03_04.pdf.

Harvey, D. 2003. *The New Imperialism.* Oxford: Oxford University Press.

Haydu, J. 1999. Counter action frames: Employer repertoires and the Union Menace in the late nineteenth century. *Social Problems*, 46(3). August: 313–331.

Hironaka, A. 2002. The globalization of environmental protection: The case of environmental impact assessment. *International Journal of Comparative Sociology*, 43(1): 65–78.

Human Rights Law Network. 2008. *Independent People's Tribunal on Dams in Arunachal Pradesh: Interim Report.* New Delhi: Human Rights Law Network.

Isaac, L. 2008. Counterframes and allegories of evil: Characterizations of labor by gilded age elites. *Work and Occupations*, 35(4): 388–421.

Iyer, R. R. 2001. World commission on dams and India: Analysis of a relationship. *Economic and Political Weekly*, 36(25). June 23–29: 2275–2281.

Kingdon-Ward, F. 1952. Caught in the Assam-Tibet earthquake. *National Geographic Magazine*, 150(3): 2–16.

Kingdon-Ward, F. 1953. The Assam earthquake of 1950. *The Geographical Journal*, 119(2). June: 169–182.

LSEG (Lower Subansiri Expert Group) Report. 2010. *Report on Downstream Impact Study of the Ongoing Subansiri Lower Hydroelectric Project at Gerukamukh of National Hydroelectric Power Corporation Limited.* Submitted by Expert Group (Jatin Kalita *et al.*), Guwahati.

McCully, P. 2001. *Silenced Rivers: The Ecology and Politics of Large Dams.* New York: St. Martin's Press.

Mol, A. 1999. 'Ontological Politics: A Word and Some Questions'. In: Law, J. and H. John (eds.). *Actor Network Theory and After*. Oxford: Blackwell, pp. 74–89.

National Hydroelectric Power Corporation (NHPC). 2010a. 'Counterpoint: The NHPC's view point'. In: Mrinal Talukdar and Kishor Kumar Kalita (eds.). *Big Dams and Assam*. Guwahati: Nanda Talukdar Foundation, pp. 33–49.

National Hydroelectric Power Corporation (NHPC). 2010b. 'Face off: The Expert Group versus NHPC'. In: Mrinal Talukdar and Kishor Kumar Kalita (eds.). *Big Dams and Assam*. Guwahati: Nanda Talukdar Foundation, pp. 50–68.

National Hydroelectric Power Corporation (NHPC). 2012. *Clearing the Myths by Presenting Facts on Subansiri Lower Hydro Power Project*. Advertisement. *Telegraph* (Guwahati). January 21.

Nixon, R. 2010. Unimagined communities: Developmental refugees, mega dams and monumental modernity. *New Formations*, 69: 62–80.

Polanyi, K. 1980 [1944]. *The Great Transformation*. New York: Octagon Books. *Public Interest*, 28, pp. 38–50.

Ramanathan, K. and P. Abeygunawardena. 2007. *Hydropower Development in India: A Sector Assessment*. Manila: Asian Development Bank.

Sahni, A. 2012. The Northeast: Troubling Externalities. *South Asia Intelligence Review, Weekly Assessments & Briefings*, 10(38). March 26. http://www.satp.org/satporgtp/sair/Archives/sair10/10_38.htm (Accessed on 6 June 2017).

Talukdar, M. and K. K. Kalita (eds.). 2010. *Big Dams and Assam*. Guwahati: Nanda Talukdar Foundation.

Thompson, E. P. 1975. *Whigs and Hunters: The Origin of the Black Act*. London: Allen Lane.

Vagholikar, N. and M. F. Ahmed. 2003. Tracking a hydel project. *Ecologist Asia*, 11(1): 25–32.

Vagholikar, N. and P. J. Das. 2010. *Damming Northeast India*. Guwahati: Action-Aid.

World Bank. 2007. *Development and Growth in Northeast India: The Natural Resources, Water, and Environment Nexus –Strategy Report*. Report No. 36397-IN, South Asia Region. Washington, DC: The World Bank.

World Bank. 2012. *The Hidden Harvests: The Global Contribution of Capture Fisheries*. Agriculture and Rural Development Series. Washington, DC: The World Bank.

World Commission on Dams (WCD). 2000. *Dams and Development: A New Framework for Decision-Making*. The Report of the World Commission of Dams. London: Earthscan.

www.downtoearth.org.in/content/assams-dam-crisis.

www.preventionweb.net/files/12782_icimodadjustingtofloodsonthebrahmap.pdf.

www.satp.org/satporgtp/sair/index.htm.

Part II

Case studies

9 Conflicts over embankments on the Jiadhal River in Dhemaji District, Assam

Partha J. Das

Introduction: setting the context

With 3.1 million ha, or some 40 per cent of the total geographical area of the State being flood prone, Assam is one of the most flood affected States of India. Floods are caused annually during the pre-monsoon and the monsoon seasons by its numerous rivers, rivulets and streams belonging mainly to two major river systems – the Brahmaputra and the Barak (Meghna). The State has experienced major floods in the years 1954, 1962, 1966, 1972, 1977, 1984, 1986, 1988, 1998, 2002, 2004, 2008, 2011 and 2012. In the aftermath of the great earthquake of 1950, the intensity, frequency and the damage due to floods increased progressively. The floods of 1988, 1998, 2004, 2008 and 2012 were the worst in recent history.

About 90 per cent of the agricultural land and urban population centres, as well as Assam's most valuable economic assets such as tea estates, oil fields, roads, and airports, are annually damaged by floods. On an average, an estimated $47 million in annual crop production is lost due to floods, while damage to homesteads and livelihoods affects some 3 million people (ADB 2006). Riverbank erosion is also a chronic problem caused by dynamic shifting of channels flowing through unconsolidated heavy sand or silt strata of the floodplain, with high sediment discharge. Between 1954 and 2010, Assam's 17 riverine districts have lost 7 per cent of their land area to erosion. Some 8,500 ha of land (valued at $20 million) is lost annually. As a result of floods and erosion, about 10,000 families are displaced (ADB 2006), many of whom become landless each year, causing significant social and economic disruption.

Flood management in the State has not been effective to the desired extent in protecting people's lives, livelihoods, property and public infrastructure, mainly because of serious shortcomings in the flood management system. Flood and river bank erosion are mitigated in the

State (like elsewhere in the country) mainly by deploying structural measures such as embankments, canals, sluices, revetments, spurs, porcupine, etc. These measures are adopted to achieve the combined purpose of containing floods, protecting river banks and embankments from erosion, and helping land formation in rivers. However, these strategies usually do not provide a long-term solution to the problem; rather, sometimes these structures become counterproductive and aggravate the problem that they are supposed to mitigate. Maintenance of the embankments is poor resulting in frequent breaching leading to devastating floods (Das and Bhuyan 2013). Embankments determine both their security from floods as well as vulnerability to floods; embankments also significantly influence people's decisions and action for adapting to floods (Das *et al.* 2009).

Although there has been discussion and promises of complementing structural mitigation strategies with nonstructural measures since the first flood policy of the country was announced in 1954 (Brahmaputra Board 1985; Mishra 2002), flood management in Assam is overwhelmingly dominated by structural measures only. Assam has about 4,459 km of embankments, 851 km of drainage channels and about 681 town/village protection works (as bank protection and anti-erosion measures) covering an area of 1.64 million ha, while 2.18 million ha of land still remained to be protected as of 2004 (MWR 2004). About 10 years hence the situation is not much different in 2014 with very few new embankment projects implemented in between.

Years of suffering from flood and erosion and consequent impoverishment have compelled the riparian communities in many areas in the State to resort to agitations demanding properly built embankments as a means of security against flood and erosion. This chapter examines the long-standing conflict between local communities and the government authorities over flood control measures, especially with respect to embankments in the Jiadhal River in the Dhemaji District of Assam (Map 9.1), which is one of the most flood affected districts of the country. The embankments and related structures often fail to protect people from floods in the Jiadhal area. Floods triggered by breaching of embankments create more devastation. Embankments are vulnerable to breaching because these are neither properly built nor regularly repaired. The study was conducted as part of a larger project on 'Governance of flood mitigation infrastructure in Assam' with support from the International Centre for Integrated Mountain Development (ICIMOD), Kathmandu in 2010–12. This case study has tried to document the most recent developments in the conflict situation in the study area up to August 2015.

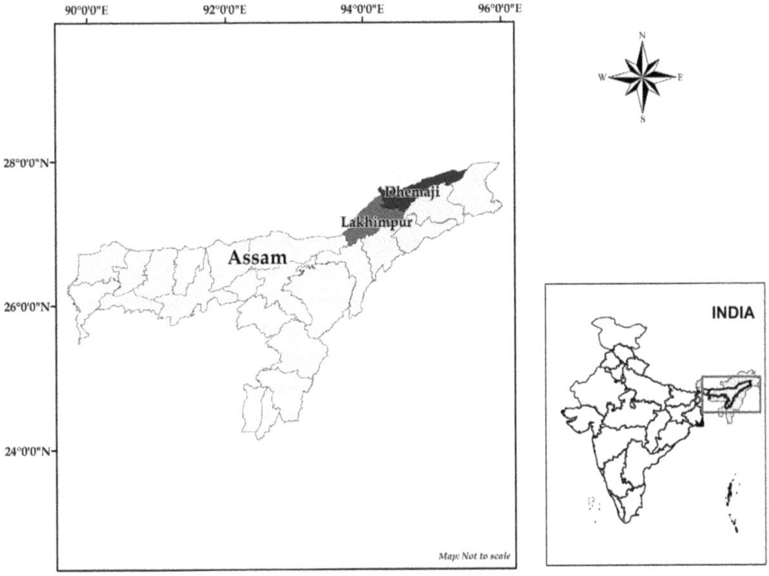

Map 9.1 Case study area, Dhemaji District of Assam
Source: SOPPECOM, Pune

Study area: the flood hazard scenario in the Jiadhal River catchment

The study area comprises the catchments of the Jiadhal River Basin in general and the flood affected areas on river banks having embankments in particular. The river Jiadhal, a north bank tributary of river Brahmaputra, originates in the lower Himalayan ranges in the West Siang District of Arunachal Pradesh and flows through the Dhemaji District of Assam (Map 9.2) in a complex network of channels, to finally merge partly with the Subansiri River, a major north bank tributary of the Brahmaputra. It has a catchment area of 1,205.41 sq. km out of which 370.41 sq. km (30.73 per cent) lies in the hills of Arunachal Pradesh and 835 sq. km (69.27 per cent) lies in the plains of Assam (Das 2012). The average annual rainfall in the catchment varies from 2,965 mm to 4,386 mm. The Jiadhal is typical of many other north bank tributaries of the Brahmaputra which generally flow in shallow braided channels, have steep slopes, carry a heavy silt charge and are flashy in character (Goswami 1998). In the plains of Assam

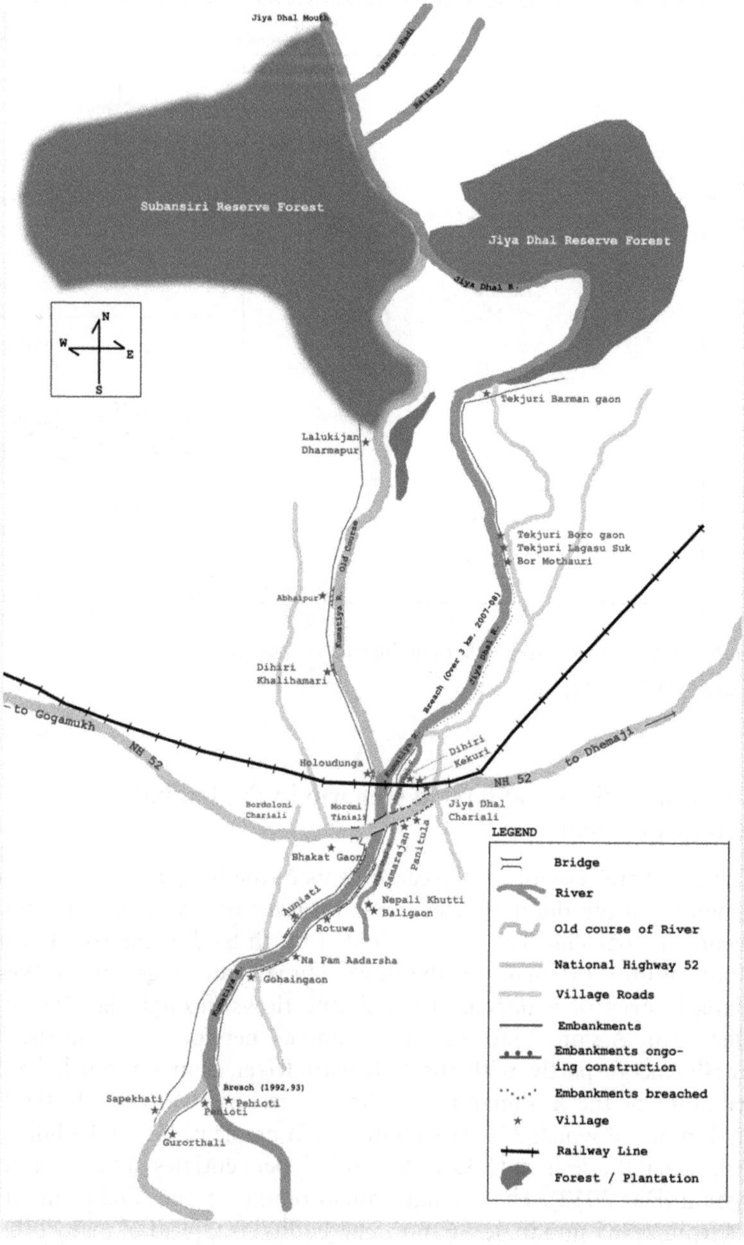

Map 9.2 Map of Kumatiya-Jiadhal River with communication, transport and flood protection infrastructure

Source: Author

the river refuses to flow in a definite course, earning a propensity and notoriety for frequently migrating in its channel and changing its course unpredictably almost in every flood season (Brahmaputra Board 2000).

Large silt yields have resulted in acute channel aggradations and a rising river bed. It is a highly dynamic river showing very high variations in annual maximum discharges and corresponding water level in a basin where flood hazard is increasing in terms of sedimentation, sand casting, flood inundation area and frequent shifting of channel (Hazarika 2003). Flooding due to bank spillage and breaching of the embankment, acute erosion of the riverbank, frequent shifting of the river channel and consequent floods as well as degradation of land due to deposition of layers of sand are chronic problems of this basin. The Jiadhal Basin experiences severe floods every year and the floods cause enormous damage to property, agriculture, land and surface communication. However, the most menacing of the water-induced hazards is the widespread deposition of sand carried by the flood on the fertile agricultural land. About 572 ha of land lying in more than a hundred villages has been seriously affected by sand casting (Deka 2010).

In recent times, devastating flash floods were recorded on the Jiadhal River in 1984, 1988, 1989, 1992, 1994, 1997, 1998, 2002, 2007, 2009, 2011 and 2014. The flood damage in 1997 in terms of money was Rs. 3.3 million (Indian Rupee) while in 2002 it was Rs. 141.23 million (Hazarika 2010). About 680 villages, 9,88,502.5 ha of land and a population of 23 million were affected by the floods of 1998. Forty-eight people died in the floods during 1989–2002 (Hazarika 2010). The major flood in 2007 caused by a failure of the right embankment near the bridge on the Kumatiya branch of the river destroyed about 30 villages on the right bank. The flood in 2009 was let out by the left embankment adding to the prolonged misery of the people. The river experienced five waves of flash floods during the rainy season in 2011, with the most catastrophic 15 August 2011 flood affecting about 85 villages severely and 300 hectares of crop land, including a large area with standing crops. Most of these flash floods resulted in failure of embankments and changes in the river course in several places, making the floods more hazardous.

The Samarajan area has become the epitome of annual flood mayhem since 1984 when the 'Basin embankment'[1] on the left bank of the river near Dihiri village breached for the first time, and the major flow of the river was discharged through a channel (now called the Jiadhal channel) through Samarajan. Since then, this branch of the river has become more powerful carrying a huge amount of flow and sediment, and the left bank of the river downstream, located to the south of the National Highway-52 (NH-52) was severely affected by floods, with new areas coming under the impact of floods and erosion as the river has been branching out breaching embankments at many places and creating a mesh of channels as it flows down. The river now flows in two major channels, the Kumatiya and the Jiadhal, with more discharge going down the Jiadhal channel during the monsoon season and as a result creating more devastation on the left bank areas.

Community response to failing flood structures: emergence of the conflict

As mentioned earlier, the major portion of flood waters from the river started flowing through the dynamic and newly created Jiadhal channel after the floods of 1984, which were triggered by the collapse of the basin embankment in Dihiri village, just upstream of the NH-52 (Figure 9.1). People living on the left bank of the river began to be disgruntled with the State and district authorities since this disaster.

Figure 9.1 Dilapidated condition of embankments on the Jiadhal River, Dhemaji District

Source: Author

The affected people started demanding proper compensation, rehabilitation and river training measures so that the river is diverted towards the west, and the bulk of it flows through the original channel of Kumatiya. The failure of government schemes to control flood and erosion, the wanton mismanagement of funds and the abysmal lack of transparency and efficiency in implementing flood mitigation projects for decades have disillusioned the people about the government's capability and commitment to mitigating flood hazards. People affected by the Jiadhal floods have been demanding proper repairing and strengthening of embankments. However, over the last 30 years, no effective interventions were even though floods have become more devastating. As a result people in the entire West Dhemaji region are highly critical of the State's Water Resources Department (WRD) because of its failure to mitigate flood and erosion in both the Jiadhal and Kumatiya channels.

The Jiadhal Nadi Baan Pratirodh Oikya Mancha (JOM),[2] which evolved as a platform for villagers from eight different panchayats affected by floods and erosion from the Jiadhal River in the last 20 years, has been consistently engaging with the WRD and district administration to expose cases of misappropriation of project funds and sloppy work. The JOM, an umbrella organisation of various community organisations and individuals, is leading the local movement to demand a permanent solution to the flood and erosion problem through properly designed and genuinely implemented structural measures. It has submitted memorandums, organised awareness meetings, rallies and sit-in protest as part of its nonviolent movement for flood security, compensation against flood damage and proper resettlement and rehabilitation (R&R). Interestingly, the bulk of the activism against poor flood management is taking place in the villages on the left bank of the Jiadhal River, whereas the influence of the campaign is relatively weak in villages on the right bank. This is possibly because the right bank has been affected less by floods (from the Kumatiya channel) than the left bank, since the Kumatiya branch carries less flood discharge at present compared to the Jiadhal branch. Moreover the demand of the JOM led people to channelise the main Jiadhal River from an upstream point towards west would make most of the water to flow through the Kumatiya channel (as it was in the pre-1984 years) is not supported by the villagers living on the right bank since they think such a manoeuvre would again make their villagers more vulnerable than at present (Figure 9.2).

Led by the JOM, people have demanded the following measures: (1) repairing and strengthening of the old Basin Embankment and

Figure 9.2 Activists of the JOM addressing the people in a meeting at Samarajan on 11 October 2011

Source: Author

its extension (including the portion that breached in 2007) further downstream at least to Nepali Khuti or about 12 km from the northernmost point where the Basin dyke begins, (2) stalling repairs of the abandoned Moridhal dyke, (3) stopping any attempt to construct an embankment that cuts through existing villages, (4) repealing the recent plan of an embankment only on the southern part of the NH-52, and (5) including people's opinion and views in planning and implementing flood mitigation structures, (6) increasing height and length of the railway bridge on Jiadhal River so that it is not affected and damaged by rising flood levels of the river due to silting of the river bed and (7) getting adequate monetary compensation of flood and erosion affected people including R&R.

For the last three decades, the community has been demanding that the original embankment, known as the Basin embankment, constructed on the left bank in the year 1972 and then extended twice in subsequent years, should be strengthened keeping the same alignment and maintained regularly. At the same time, anti-erosion measures should be deployed on the left bank to prevent loss of land in the villages. They want this embankment to be built strongly from the foothills in the north to the areas far downstream like Nepali Khuti, with state-of-the-art technology, so that the river cannot erode the left bank and make its way into their settlement and agricultural areas any more. They also want the government to push the river to the west towards the original course (through the Kumatiya channel), by taking river training measures in the upstream section so that the river can flow through a definite channel and confined course bound by embankments on both banks (Figure 9.3).

The communities feel that instead of paying heed to the people's opinions and demands, the State WRD was wasting money and time every year on repairing the old Moridhal embankment that runs from Tinigharia through Kekuri and Dihiri villages (where embankments breached in 1984 and then frequently thereafter) to the railway line at Samarajan. This old and dilapidated dyke has never been repaired well, and it has perpetuated for decades without providing succour to

Figure 9.3 Local villagers attending a meeting of the JOM to discuss strategies of their movement seeking protection against flood and erosion

Source: Author

the people. The community sees vested interest of officials and contractors in retaining this old useless structure for making money in the name of maintenance.

Police forces were deployed to dilute people's resistance in the month of April 1989, when the local villagers formed a barricade at the site to stop the WRD from doing repair work on the old Moridhal dyke. The villagers demanded proper maintenance of the extension parts of the Basin embankment instead. They were beaten up by the police, and those who incurred serious injuries included children, old men and women. On 16 November 1994 picketers were beaten up by the police in front of the office of the Deputy Commissioner of Dhemaji, when they were protesting against the ongoing technical deficiencies and monetary corruption in the embankment projects.

Contrary to people's wishes and demands, the WRD then tried to build a new embankment through the Dihiri village in 2009, following an alignment about a kilometre away from the riverbank on the plea that the P. P. Verma Committee, constituted by the Government of Assam to look into the flood and erosion problem of Dhemaji District, had suggested the construction of an embankment at a safe distance from the riverbanks.[3] The community protested vehemently against the construction because this embankment would leave about four villages and a number of small settlements outside exposed to the river devoid of any protection. They stalled the construction on several occasions. Intervention of the local Member of Legislative Assembly (MLA) of Dhemaji favouring the government's decision was also thwarted by the people in March 2009 leading to a stalemate.

During this stalemate, the WRD built four more spurs before the rainy season began in 2011, using boulders and geo-bags in the upper stretch of the river near Tekjuri Barman Gaon in order to push the river towards the west and save the left bank villages from erosion. However, JOM activists and many other villagers of the area were sceptical from the word go about the effectiveness of these measures due to the poor quality of construction and inappropriate location of the spurs. That the villagers were right in their understanding was vindicated when the spurs were swept away by several waves of flash floods which devastated the Jiadhal area during July, August and September in the year 2011.

In the aftermath of the big flash flood on 15 August 2011, the WRD proposed to construct another embankment on the left bank south of the NH-52 to protect the villages of West Dhemaji from floods. The community, especially those from the north of the NH-52, united under the JOM, resented this decision because in their opinion if the upper reaches of the river (north of NH-52) are not protected with strong embankments, any attempt to protect the lower reaches (south of NH-52) will certainly fail.

The community submitted a memorandum to WRD officials, the district administration and the State Water Resources minister reiterating their demand for the confinement of the river in a single channel by robustly built embankments close to existing riverbanks so that a minimum area of their settlement and agricultural land is left outside the embankments. They are also urging the authorities to forego the plan to construct a new embankment only on the southern part of the left bank leaving the northern (upstream) stretch unprotected. Meanwhile, from September to December in 2011, the activists of the JOM reconstituted the organisation and strengthened their mass base and network in a bid to prepare themselves for a new phase of the movement.

The stalemate and conflicts continued while the JOM tried to consolidate its strategy and strength through awareness campaigns. On 26 November 2013 they resorted to road barricade on the NH-52 near Samarajan and stopped transportation of vehicles about for 6 hours repeating the same demands. About 500 agitators led by JOM and supported by many local organisations participated in this protest programme.

The Jiadhal River just down of erstwhile Tekjuri Barman Gaon breached the dyke, shifted its channel and flew through the Bhaju area on 21 August 2014 night. About 15 villages were affected severely by floods and erosion during this flood including about 1,500 Ha of cultivated land and 10,000 people (WRD 2014). In a usual manner the WRD came up with a plan to plug this breach and divert the river to the original course (before 22 August 2014) by channel cutting and construction of a coffer dam near the point of breaching. The local communities led by the JOM suggested another place in upstream as the location where the diversion scheme should be implemented in order to divert a major chunk of the water to flow through the Kumatiya channel.

The flood activists raised this and the other core issues with the Deputy Commissioner and the local MLA. But the State WRD went on and executed the work in their own way on the pretext that what the local people wanted them to do would have been expensive. The locals are not happy with what was done in the name of plugging the breach; they believe much more flow could have been diverted trough the Kumatiya if had the WRD listened to their suggestion. Moreover there is now a possibility that the river in the event of a major flood may hit the same dyke again and cause another breach at the same place.

The JOM organised a People' Convention on the Jiadhal issue on 30 November 2014 and its representatives met the Chief Minister of Assam at Guwahati and submitted memorandum with recommendations of the Convention on 10 January 2015. They also organised another road barricade on NH-52 near Jiadhal Chariali on 2 April 2015 with the same demands and, submitted memorandum to Chief Minister through deputy commissioner.

Box 9.1 and Box 9.2 present the chronology of major events of failure of flood mitigation structures and milestones in the 30-year long history of the conflict.

Box 9.1 Chronology of the collapse of embankments

- May 1984: The old Moridhal embankment breaches in Dihiri village; the Samarajan area is devastated.
- 13 May 1988: Extension phase-I in Basin embankment and old Moridhal dyke collapses.
- 22 June 1992: The old Moridhal embankment breaches in Kekuri village.
- 7 July 2007: Kumatiya dyke on the right bank of the Kumatiya branch of Jiadhal breaches near Dihingia village in Bordoloni Development Block, causing an unprecedented flood havoc in more than 50 villages on the right bank.
- 13 August 2009: Embankment on the left bank of Jiadhal south of NH-52 breaks downs in the wee hours at Nepali Pathar and Adarsha Gaon; heavy flash floods in Jiadhal and Kumatiya; the NH-52 in Samarajan is breached; there is widespread devastation on both banks.
- 15 August 2011: Catastrophic flash floods in Jiadhal; almost the entire Jiadhal catchment is affected; about 85 villages are affected seriously; the river creates a new channel, and engulfs Tekjuri Barman Gaon and Dihiri.
- 21 August 2014: The Jiadhal River just down of erstwhile Tekjuri Barman Gaon breached the dyke on the left bank, changed its course and flew through the Bhaju area.

Box 9.2 Chronology of conflicts

- 10 May 1989: Agitators who prevented the WRD from repairing the Moridhal dyke near Dihiri demanding better and holistic maintenance of embankments are beaten up by the police; children, old men and about 19 women are injured.
- 30 March 1994: A group of seven representatives from the flood affected areas of Jiadhal observed fasting for eight days at Guwahati.
- 11 November 1994: Picketers protesting against corruption in the embankment projects are beaten up in Dhemaji, 23 people are injured.
- 22 December 2011: JOM submits a memorandum to the Water Resources Minister of Assam at Guwahati.
- 29 December 2011: The Water Resources Minister visits areas critically affected by flood and erosion on the left bank of Jiadhal River.
- 26 November 2012: Local people led by JOM and supported by many local organisations resorted to road barricade on the NH-52 near Samarajan and stopped transportation of vehicles about for six hours.
- 30 November 2014: People's Convention on the Jiadhal Flood & Erosion Problem in Jiadhal College, Dhemaji.
- 10 January 2015: Representatives met the Assam Chief Minister at Guwahati and submitted memorandum.
- 2 April 2015: Road barricade on NH-52 near Jiadhal Chariali on same demands, followed by submission of memorandum to Chief Minister through DC.

The JOM has also developed a plan of embankments with schematic diagrams incorporating their learning and experience of the impacts of structural measures, and local knowledge of the river's behaviour. Some of the villagers involved with the movement are a storehouse of knowledge about the river, and their expertise in predicting the behaviour of the river is so well established that people call them barefoot engineers. In the community plans they have highlighted the loopholes in the official schemes of embankments and related structures. Suggestions have been offered to the WRD and the district administration towards making more efficient and durable structures as well as containing the river in a defined channel. However, the departmental engineers are reluctant to recognise the knowledge and wisdom in the community plans, and refuse to give credence to the community's opinion demands.[4]

According to the people, the sites of construction of a coffer dam and a flood wall (WRD 2007a) were wrongly selected by the WRD engineers during 2007, because of which the structures didn't survive the floods that year. Although the right locations were pointed out by the people, they were ignored on the plea that the coffer dam and river diversion project were approved by the Central Water Commission (CWC) after model testing (WRD 2007b). Committees were formed in some villages at the request of the WRD to mobilise public opinion about the WRD's schemes and incorporate suggestions of the people for better schemes, but the committees were never consulted while the scheme was executed.[5]

Although the villagers of the Jiadhal area have been agitating peacefully for the last two decades, now led by the JOM, other civil society organisations have also carried out protest programmes in demand of a sustainable solution to flood and erosion problem in the district. Prominent among these are the Asom Jatiyatabadi Yuba Chatra Parishad (AJYCP),[6] the All Assam Students' Union (AASU),[7] Takam Mising Porin Kebang (TMPK),[8] Krishak Mukti Sangram Samiti[9] (KMSS), and the Dhemaji Zila Ban Samaishya Sthayi Samadhan Dabi Mancha.[10]

Flawed governance: cause of poor flood mitigation and conflicts

The responsibility of flood protection in the State lies mainly with the WRD. There are several major shortcomings in the institutional norms and practices that are followed by the WRD in decision making about structural interventions and their execution. The Technical Advisory Committee (TAC), the apex decision-making body on technical matters, consists solely of members from government departments and agencies, and that too mostly engineers. This explains why our flood management schemes are always dominated by engineering and structural interventions without taking into consideration the environmental and social impacts and risk assessment. Even local governance agencies like the Panchayati Raj Institutions (PRIs) and Autonomous Councils are rarely consulted while planning and approving the structures. Thus, there is no practice of integrating local people's opinions and indigenous wisdom in the embankment schemes (Das and Bhuyan 2013).

Fund crunch is cited as the main reason for improper maintenance of the embankments. However, it is not the only reason. Misuse of funds and lack of efficiency of engineers and officials add to the failure of flood control measures. The prevailing budgetary cycle also hinders timely completion of construction and maintenance of embankments. By the time funds are released for repairing or construction of embankments in March and April, the rainy season sets in and early floods start occurring. As a result, the work becomes more expensive, and the quality of the work cannot be maintained within the given timeframe with limited funds. Such a mismatch between the monsoon calendar and the budget cycle renders even genuine efforts of a few well-meaning officials and contractors ineffective.

Inefficiency of the flood management approach, misappropriation of funds, mistiming of resource allocation, and frequent failure of flood mitigation structures, the four main causes that lead to conflicts between people and the government, are essentially problems of governance. Institutions created for flood management, both at the Central and State levels, do not show the adequate commitment, efficiency, transparency, accountability, flexibility, innovation and vision needed for effectively managing these water induced hazards. There is poor implementation of the existing policies at the State and district levels. Also, in many cases, policies are inadequate or nonexistent when it comes to ensuring good governance of flood management, to make it transparent, participatory, equitable, technically efficient and socially just. As a result, at the State and district levels, decision making in official flood management is a top down, opaque, and rigid process practically detached from participation of communities and civil society.

The political economy of flood control programmes especially that of embankments play a major role in creating vested interest groups, often a nexus of officials, contractors and politicians which want to perpetuate the problem of flood and erosion. They act in favour of endorsement of structures that have been badly constructed, get projects sanctioned in places that are inappropriate, and ignore places which need genuine flood protection. They want to make sure that embankments fail regularly so that their annual maintenance and repairing becomes an economically profitable venture for the nexus. After struggling for 30 years the people of the Jiadhal banks have started doubting the intention of the concerned authorities. They are asking questions whether the State Government has any genuine interest in long-term solution of the flood and erosion problem. Or a powerful lobby in the State wants keep the problems alive so that the business of piece-meal and short-term flood mitigation measures thrive on for their illegitimate economic benefit.

The way forward

It is high time that the government reforms the relevant institutions, policies, legislations and practices in order to implement holistic flood and erosion management. The prevailing paradigm of 'controlling floods' should give way to 'mitigating flood impacts', with an participatory approach that will ensure application of only sensible and flexible river engineering along with empowering people to become more adaptive and resilient to flood risks. The WRD needs to become more flexible, transparent and participatory in its functioning. The government will do well to adopt a comprehensive State-level policy for flood and river erosion management with adequate provisions for guidelines for construction and maintenance of flood mitigation structures as well as nonstructural measures. To resolve the conflict, the government must recognise the community's demand for a right to participate in the decision-making process of flood mitigation.

Although it is important to recognise and incorporate the community's knowledge and opinions in official flood protection work, one needs to be careful about implementing technical schemes wholly based on community recommendations, especially in the case of an unstable and dynamic river like the Jiadhal. According to D. K. Deka, former Water Resources Secretary of the State, 'let the planning of technical interventions be enriched with indigenous wisdom, but leave their implementation only to water resources engineers'.[11]

Similarly, the overwhelming emphasis on construction of embankments and river engineering seen in the community plans and charters may not be fruitful in all situations. While the community's knowledge and their aspiration for participation in the flood mitigation process must be respected and endorsed, at the same time, there is a need to sensitise the people about an integrated approach to flood and erosion hazard mitigation based on lessons learned from the research and application of technology from all over the world.

There are also conflicts of interest between the right and left bank dwellers in their campaign for flood protection. The communities fighting for flood security under the aegis of the JOM are mostly from the left bank villages. They want the main course of the river to be pushed towards the west (right bank) to make the river flow through the Kumatiya channel (the original main stream before 1984). However, it may not be a desirable and sustainable solution as such a step may be resisted by people living on the right bank, who are now less affected by floods. Although now dormant the conflicts of interest between the people living on the left and the right banks of the Jiadhal River may come out openly in the near future. Therefore the JOM needs to be more inclusive in its mobilisation of community opinions so that people living on both banks are equitably benefitted.

Notes

1 The oldest embankment on the Jiadhal river, constructed in 1972 from the northern foothills to near the railway line.
2 'United Platform for Resisting Floods of Jiadhal River', the leading movement protesting against of the State Government's inactions and wrong actions in flood management specific to the Jiadhal river.
3 Interview with Harish Pegu, a leading activist of the JOM, on 12 December 2010 at Samarajan (Dhemaji).
4 Interview with Gandheswar Bora, a leader of the Jiadhal flood justice movement, at Barman Gaon on 19 January 2010.
5 Sourced from Focus Group Discussion (FGD) at Dihiri Mising Gaon, Jiadhal Panchayat on 18 March 2010.
6 AJYCP is a frontline organisation of students and youth in the state which has led movements on various social and political issues such as large river dams, flood and erosion, total autonomy for Assam, etc.
7 AASU is the largest students' organisation pursuing issues like illegal immigration, flood and erosion, and large river dams critical for the survival of indigenous people in Assam. It led the historic Assam Movement in the 1980s, and is engaged in negotiations with the Government of India on matters related to the Assam Accord.
8 TMPK is the platform of the students and youth of the Mising community.
9 'The committee for struggle for liberation of farmers' is a grassroots organisation leading a social movement against the hydropower projects in North East India and for land rights of the indigenous people of Assam.
10 'Platform for demanding permanent solution to the flood problem of Dhemaji District', is a recently formed organisation of citizens of Dhemaji district.
11 Interview with D. K. Deka, former Water Resources Secretary of the state, at Guwahati on 14 July 2010.

Reference

ADB. 2006. *India: Preparing the North Eastern Integrated Flood and Riverbank Erosion Management Project (Assam)*. Technical Assistance Report. Project Number: 38412. Asian Development Bank. December 2006. www.adb.org/Documents/TARs/IND/38412-IND-TAR.pdf.

Brahmaputra Board. 1985. *Master Plan of Brahmaputra Basin*. Part I. Main Stem. Volume I. Brahmaputra Board. New Delhi: Ministry of Water Resources. Government of India, pp. iv–3.

Brahmaputra Board. 2000. *Master Plan of Jiadhal Sub-Basin* (A Master Plan under part-III). Brahmaputra Board. New Delhi: Ministry of Water Resources, Government of India.

Das, P. J. 2012. *Building Community's Capacity for Flash Flood Risk Management in the Jiadhal River Basin in Dhemaji District, Assam, India.* Technical Report Submitted by Aaranyak to the International Centre for Integrated Mountain Development (ICIMOD). Guwahati: Unpublished.

Das, P. J. and H. K. Bhuyan. 2013. *Policy and Institutions in Adaptation to Climate Change: Case Study on Flood Mitigation Infrastructure in India and Nepal.* ICIMOD Working Paper 2013/4. Kathmandu: ICIMOD. http://lib.icimod.org/record/28382/files/WP_4_13.pdf.

Das, P. J., D. Chutiya and N. Hazarika. 2009. *Adjusting to Floods on the Brahmaputra Plains, Assam, India.* Kathmandu: The International Centre for Integrated Mountain Development (ICIMOD). http://lib.icimod.org/record/8025/files/attachment_669.ps.gz.

Deka, C. R. 2010. *Land Degradation by Jiadhal River of Dhemaji District: A Study Using Remote Sensing and GIS Technique.* Assam: ENVIS. October–December 2008. www.envisassam.nic.in.

Goswami, D. C. 1998. Fluvial regime and flood hydrology of the Brahmaputra River, Assam. *Memoir Geological Society of India*, 41: 53–75.

Hazarika, U. M. 2003. Problems of Flood, Erosion and Sedimentation in the Jiadhal River Basin, Dhemaji District, Assam: A Geo-Environmental Study. Unpublished PhD thesis submitted to Gauhati University.

Hazarika, U. M. 2010. Fluvial environment of Jiadhal river basin, Dhemaji district, Assam. *International Journal of Ecology and Environmental Sciences*, 36(4): 271–275.

Mishra, D. K. 2002. *Living with the Politics of Floods: The Mystery of Flood Control.* Dehradun: People's Science Institute, p. 124.

MWR. 2004. *Report of the Task Force for Flood Management and Erosion Control.* New Delhi: Ministry of Water Resources, Government of India. December 2004, p. 135.

WRD. 2007a. *Detailed Estimate for the Project 'Controlling of Jiadhal River in Dhemaji District, Phase-I (Recast)'.* Water Resources Division, Dhemaji. Water Resources Department, Guwahati: Government of Assam.

WRD. 2007b. *Detailed Estimate of the Scheme 'Immediate Measures for Breach Closing Works to the Controlling of Jiadhal River in Dhemaji District', Phase-I at 11th to 14th km for 2007–08.* Water Resources Division, Dhemaji. Water Resources Department, Guwahati: Government of Assam.

WRD. 2014. *Immediate Measures to Jiadhal L/B Embankment from Basin to Railway Line for 2014–2015 (Breach Closing at Barman Gaon) under SDRI.* Dhemaji: Dhemaji Water Resources Division, Dhemaji, Assam.

10 Riverbank erosion in Rohmoria

Impact, conflict and people's struggle

Siddhartha Kumar Lahiri

In the genesis of a river valley, changing river-bank-lines on the surface relief are some of the commonly observed features. However, when a major river like the Brahmaputra erodes both the bank-lines, and its channel-belt expands within a relatively short period of 90 years, the situation warrants many questions: Why is the channel belt expanding so rapidly? What causes some channels of the river to change their course so drastically? Why the Brahmaputra, which originates from the confluence of three great rivers – the Lohit, the Dibang and the Siang – is showing massive bankline shifting along both the right and left banks in the downstream direction? Considering century-scale events, the Rohmoria area in the Dibrugarh District of Assam has witnessed unprecedented riverbank erosion, which probably began in the aftermath of the great Assam earthquake of magnitude 8.7 on the Richter scale on 15 August 1950, with 532 recorded fatalities. About three years before this incident, on 29 July 1947, another earthquake of magnitude 7.9 shook the place violently. Earthquakes are an indication of the instability of a place due to deep rooted subsurface causes related to regional tectonics. Interestingly, the epicentres of both these earthquakes were only 150 km away from Rohmoria. Among natural disasters, riverbank erosion apparently seems to cause less damage than earthquakes, tsunamis, landslides or floods, because it does not claim lives usually. However, in an agrarian society, the inevitability of losing a piece of land, the most precious, and often the only possession of farmers, and the very essence of their existence, has a deep impact: a lingering trail of fear that permeates the psychological fabric of the affected community.

The objective of this chapter is to understand the nature of riverbank erosion in Rohmoria and the growing discontent of the affected people, through measurements in the geographic information system (GIS) environment complemented by frequent field visits. The ethos and trends of the people's struggle were followed as closely as possible and different interconnected issues were synthesised to comprehend the issue.

The place Rohmoria and the backdrop of the study

Geographically, as can be seen in the topographic map of the Survey of India prepared during 1912–26 (scale 1: 253,440, that is, 1 inch = 4 miles), the large village located around 25 km away from Dibrugarh town towards the upstream (southward) direction of the Brahmaputra River was known as Rohmoria (Map 10.1). With unabated bank erosion, the most affected adjoining areas in the downstream direction from Nogaghuli (where the Dibru River used to meet the Brahmaputra earlier) and up to the Balijan tea garden further upstream, were being associated with the name Rohmoria. In the first decade of the twenty-first century, Rohmoria became known as the place affected by the fastest riverbank erosion in the entire Brahmaputra Valley (Kotoky *et al.* 2005; Sarma 2008).

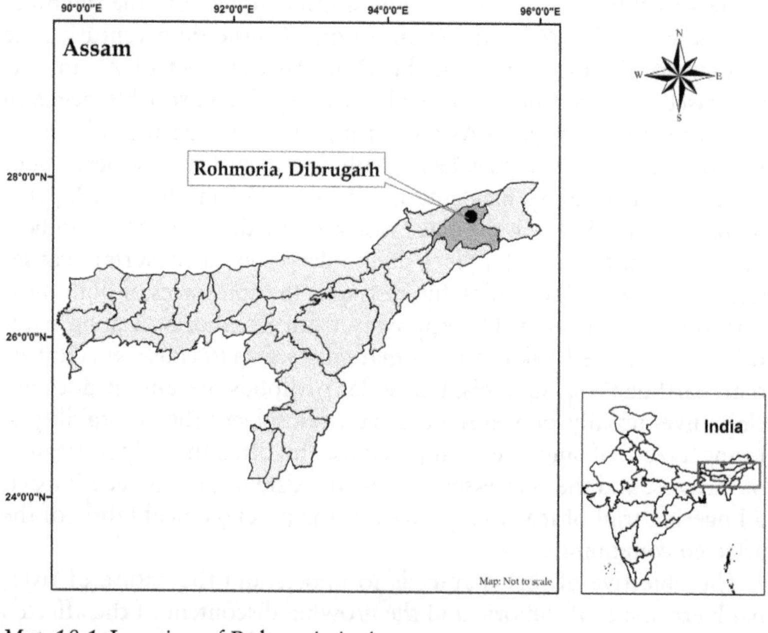

Map 10.1 Location of Rohmoria in Assam

Source: SOPPECOM, Pune

Structurally, the upper reach of the Brahmaputra Valley of which Rohmoria constitutes a small part, is sandwiched between the north-east trending Eastern Himalayas and the southern flank of the Naga-Patkai hills (Map 10.2). Some of the important thrust belts around this area are the Himalayan Frontal Thrust (HFT), the Main Boundary Thrust (MBT), the Mishmi Thrust, the Tidding Suture, the Lohit Thrust Complex (Narula *et al.* 2000) and the Naga-Patkai Thrust Belt. These thrust belts are believed to be highly active and retain the 'stretch marks' of the ancient as well as ongoing orogeny, the processes of mountain building.

Before we focus specifically on the Rohmoria area, it will be useful to remember a few broad based regional aspects related to the relationship between the valley and the mountain range, and the behaviour of the Brahmaputra River. On a global scale, the sediment yield of the Brahmaputra (Goswami 1985) within the Indian territory alone (804 tons/sq. km every year) is approximately 5.5 times higher than that of the Amazon (≈ 146 tons/sq. km every year), the largest in the world in terms of catchment area, number of tributaries, and volume of water discharged. The catchment processes responsible

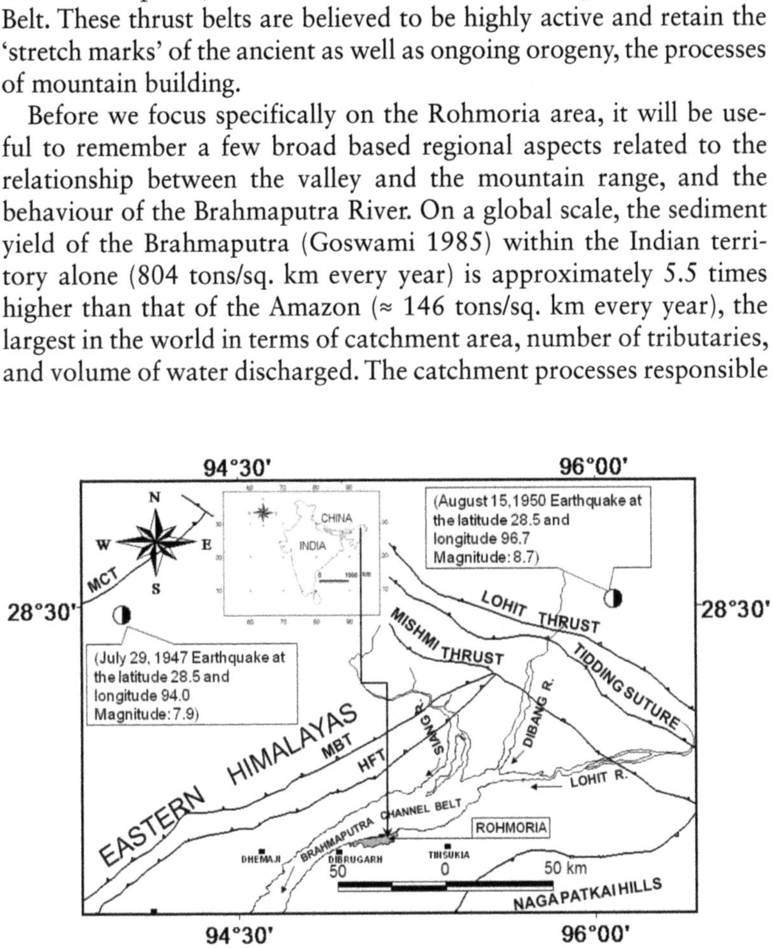

Map 10.2 Location map of Rohmoria in the backdrop of some of the important structural elements which are believed to play a fundamental role in shaping the valley. The locations of two major earthquakes are also shown.

Source: Author

for supplying vast quantities of sediment include erosion of actively uplifting mountains of the Himalayas, slope erosion of the Himalayan foothills, and movement of alluvial deposits stored in the Assam valley (Thorne *et al.* 1993). Besides this, sediment supply from the Naga-Patkai hills also contributes substantially to the ongoing sediment architecture of the valley. The drastic structural consequences (Molnar and Tapponnier 1975) in the marginal areas where the valley meets the mountain are also believed to affect the central portion of the valley, which is observed in the form of rapid changes in river dynamics.

A study of the Dibru-Saikhoa segment (Lahiri and Sinha 2012), the uppermost portion of the Brahmaputra channel-belt from the origin of the Brahmaputra River to the original confluence of the Dibru River with the Brahmaputra that includes the Rohmoria area, shows some interesting observations (Table 10.1). This segment experienced about 90 per cent positive stretching in 1975 compared to its status during 1915. The same tendency reached a whopping 230 per cent in 2005 with respect to the base year 1915. The ratio of actual channel areas and the channel-belt areas witnessed a fall, whereas this ratio

Table 10.1 Fluvial dynamics and changing landscape

Segment	*Geomorphological parameters*	*1915*	*1975*	*2005*
Dibru-Saikhoa	Average channel-belt width (in km)	5.3	10.7 (+102%)	18.5 (+250%)
	Channel-belt area (in sq. km)	358.7	678.4 (+89%)	1186.3 (+230%)
	Channel area (in sq. km)	122.0	144.1 (+18%)	270.2 (+121%)
	Interfluve area (in sq. km)	234.0	534.3 (+128%)	916. (+291%)
	Channel area/ channel-belt area	0.35	0.21 (–40%)	0.23 (–34%)
	Interfluve area/ channel area	1.92	3.7 (+93%)	3.4 (+76%)

Source: Lahiri and Sinha (2012)

Temporal variation in the parameters like average channel-belt widths, areas of the channel belts, channels, interfluves and the ratios of channel and channel-belt areas as well as interfluve areas and channel areas are shown for the Dibru-Saikhoa segment of the upper reach of the Brahmaputra Valley during 1915, 1977 and 2005. Positive numbers inside brackets indicate the growth over the 1915 values' status.

for interfluves rose greatly. This led to a very high rate of sediment dumping in the catchment areas, which could happen even if the average annual rainfall remained constant or increased slightly. The term interfluve includes all the land areas within the channel-belt. It means that later deposits like sandbars as well as earlier open lands are now encircled by channels and thereby these have become Relic islands (the anastomosing effect). The Dibru-Saikhoa Reserve Forest is such a place having an area of approximately 300 sq. km, which has appeared in the map as an island only very recently, after 1998.

Thus, the tendency of the channel-belt to increase in width caused rapid erosion and bankline migration, bringing the Brahmaputra River closer to Rohmoria. A major cause of the widening of the Brahmaputra channel-belt is the avulsion of the Lohit River through the Ananta nallah, which in turn gave rise to the formation of the Dibru-Saikhoa island. This is an example of relic island where the age of the riverine island is older than the age of the river in its present location. Another example of this type of island in the upper reach of the Brahmaputra Valley is the famous Majuli Island (Lahiri and Sinha 2014). Dibru-Saikhoa island is also known as the 'new Majuli' (Lahiri and Sinha 2015). The drastic shift of the Lohit River from its earlier course along the western side of the Dibru-Saikhoa Reserve Forest to its eastern flank resulted in a further shift of the confluence to a location closer to Rohmoria. According to older topographic maps, the confluence point used to be located near a place called Kobo in 1915. It had shifted by around 16 km in the downstream direction to a location near Laikaghat by 1975. By 2005, a further downward shift of 19 km brought the confluence point close to the upper tip of the Rohmoria area as observed in the satellite imagery. This increased the pressure on the bank at Rohmoria greatly. Further, the intrachannel belt flow characteristics continue to change, with the major channel often becoming the minor one, and vice versa. This is due to the unevenness of the sediment dispersal (Lahiri and Sinha 2012) as well as due to the locally raised flow diversion structures built by the State departments. Presently, an advancing frontier of the main channel, in the form of a 'bow' is directed towards Rohmoria (Map 10.3) (Lahiri and Borgohain 2011), making it highly erosion prone. The fourth factor is the presence of very loose sands belonging to the older floodplain deposits, just below the clayey topsoil. This facilitates rapid toe-cutting and consequent slumping of the bank. Visibly, these are the four major factors responsible for the rapid bank erosion at Rohmoria.

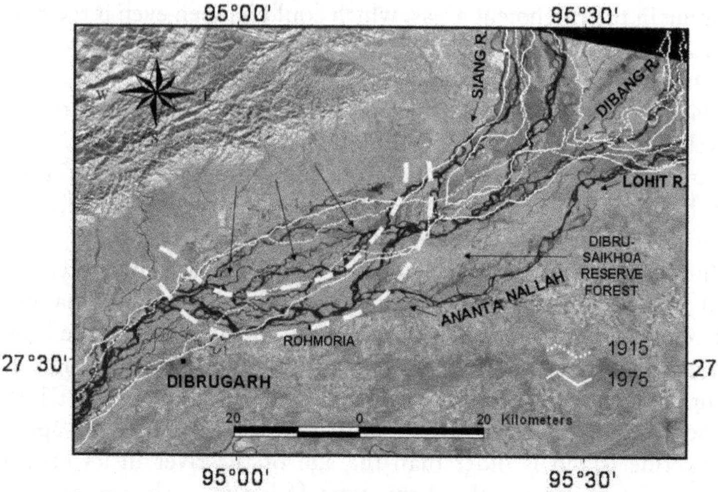

Map 10.3 A comparative study of the Brahmaputra channel-belt from 1915 to 2005 in the Dibru-Saikhoa segment of the upper Assam reach. The older channel-belts (1915 and 1975) are drawn on IRS-P6-LISS-3 image, taken on 15 December 2005 after undergoing proper georeferencing. A bow-like morphological change is distinctly observed in the main channel of the Brahmaputra River, which was nearly straight earlier. The diverging nature of the bow brings new land areas within the fold of the river by causing massive erosion.

Source: Lahiri and Borgohain (2011)

A rapid reduction in the area of Rohmoria

A comparison of the oldest topographic map mentioned above with the topographic map of 1976 (scale 1: 2,50,000) and a 23 m resolution IRS-P6-LISS-3 image, taken on 15 December 2005 provides alarming statistics. Rohmoria reduced in area from 236.54 sq. km in 1915 to 160.53 sq. km in 1975. If we assume that the bank was eroded consistently throughout this period, the erosion rate would amount to 1.27 sq. km per year.

There used to be a road called Tamuli Ali – a very old one believed to be constructed during the days of the Ahom dynasty – which connected Dibrugarh with Tinsukia town. Rohmoria lay in between. Later on, this road was renamed as the DRT (Dibrugarh-Rongagora-Tinsukia) Road. The old pre-1950 earthquake map clearly shows that the DRT Road, running more or less parallel to the Dibru River, divided the

Rohmoria area into two halves. In the 1976 map, the rapidly advancing frontier of the Brahmaputra River had already engulfed a portion of the DRT Road (Map 10.4). It is from this point onwards that the people of Rohmoria started organising themselves.

The Rohmoria struggle and the State response

The dissatisfaction of the people about the lack of action by the administration to check erosion led to the formation of an amorphous body called the Rohmoria-Lahoal-Rongagorah-Bokdung Baan Protirodh Samity in 1979. During 1979–85 a slow process of submitting memoranda to the district authority, sending delegations to meet the concerned minister, visits to the affected area by some ministers, election time assurances, etc. unfolded. The period 1985–97 was a phase of people's disillusionment (Lahiri 2008) about the non-action of the State departments. In the month of September 1997, at a big public meeting, the earlier body was dissolved and a new body named the Rohmoria Khahaniya-O-Baan Protirodh Samity was formed. This body was much stronger in its determination, and resolved to undertake some anti-erosion measures. Mobilising the people of the highly erosion prone areas belonging to 15 villages and three tea gardens, the voluntary body took up the task of constructing six wooden spurs to divert the course of the Brahmaputra River. Some technical know-how was provided by the erstwhile Embankment & Development Department of the State government. Apart from this, the entire work was based on local resources and voluntary labour. It took more than four months to complete the work. However, during the peak flood of 1998, all the spurs gave way and practically nothing remained. The setback was immense. Coincidentally, in 1998 itself, Oil India Limited (OIL) discovered a big oilfield at Khagorijan within the Rohmoria area. An oil blockade started from 16 August 1999. This day onwards, the struggle took up a political character. On 23 January 2000, a State-level organisation called the Sadau-Asom-Baan-Khahoniya-Protirodh-Sangram-Mancha was formed at the end of a three day workshop by the representatives of different voluntary organisations all over the Brahmaputra Valley, who were fighting the menace of flood and bank erosion. On 31 March 2000, Bolo Gohain, a school teacher, died under the heap of slumped soil while raising a spur. He was the first martyr for the cause of resisting the flood and bank erosion.

On the basis of the report submitted by R. A. Oak, Senior Research Officer, Central Power and Water Research Station (CWPRS), Pune to find a long term solution to the problem (site visit on 9 August 2000),

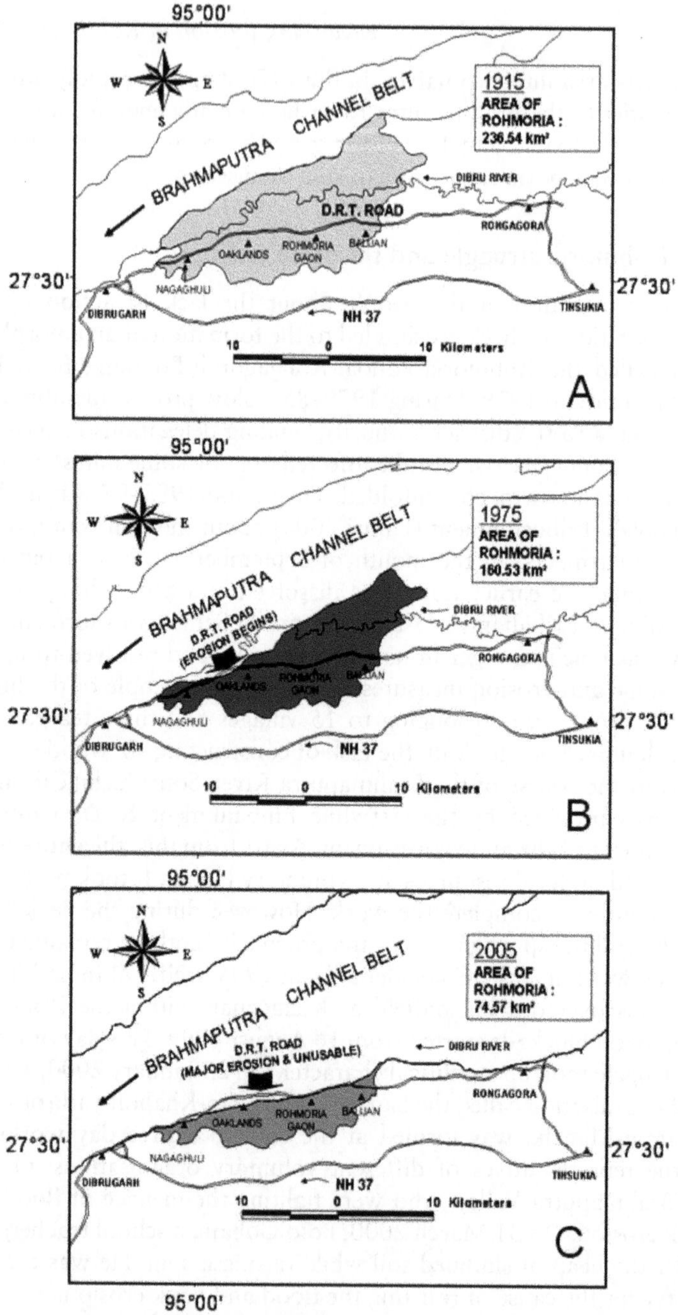

Map 10.4 The temporal change in Rohmoria, identified from the old maps and the recent satellite images. The DRT Road which earlier maintained the connectivity of Rohmoria with Dibrugarh and Tinsukia has practically vanished from the present day landscape.

Source: Author

the 31st Technical Advisory Committee meeting of the State Water Resources Department recommended a project with an estimated cost of Rs. 404.72 crore to the Central Water Commission (CWC), New Delhi for necessary approval and clearance on 24 September 2002. This included nine concrete spurs, raising of the tie-bund, and a provision of roller-compacted concrete (RCC) porcupine, along with other palliative measures. In October 2003, ten iron pipe based dampeners were raised by OIL as a pilot project. On 3 January 2004, the Dibrugarh District Administration convened a meeting of the representatives of different organisations as well as the OIL. A consensus was reached that the OIL would raise 360 iron pipe based dampeners within a stretch of 9 km (from the Oakland Tea Estate to Bogoritoli village) in a phased manner. Based on this assurance, the four and half years old oil blockade was lifted. Prime Minister Manmohan Singh visited Rohmoria on 17 January 2006. Within the period April–August 2007, OIL raised 24 more scrap oil pipe based dampeners with the help of the plan implementing agency District Rural Development Agency (DRDA), Dibrugarh. On 4 September 2007, the oil blockade was lifted again based upon the assurance that the implementation of the promised plan would be expedited. On 13 November 2007, the Central Water Resources Minister Saifuddin Soz visited the erosion affected places and declared that the Rohmoria issue would be incorporated in the next five year plan. On 20 December 2007, the Central Reserve Police Force (CRPF) teargassed and *lathi* charged about 1,500 protesters who went for the National Highway-37 blockade near the Chabua Airport Base. On 24 December 2007, a one day long Dibrugarh District *bandh* call was announced by the Rohmoria-Gora-Khahoniya-Protirodhi-Mancha against the CRPF atrocities. Other participants of the *bandh* were: All Assam Students' Union (AASU), All Assam Tea Students' Association (AATSA), Asom Jatiyatabadi Yuva Chatra Parishad (AJYCP), All Assam Tai Ahom Students' Union (AATASU) and the Motok Yuva Chatra Sammilan. In the aftermath of the flood of 2008, the Rohmoria area experienced dramatic bank erosion. On 8 September 2008, a procession of 6,000 participants with a much broader representation from many more organisations submitted a memorandum to the Prime Minister and the President of India through the district collector. The 45th Technical Advisory Committee (TAC) meeting held on 28–29 November 2008 recommended another scheme which would cost Rs. 292.048 crore based on the pre-feasibility report of the Brahmaputra Board and the model study report on the Hatighuli to Nagaghuli-Maijan reach prepared by the CWPRS, Pune. This was approved by the Secretary, Water Resources

Department, Govt. of Assam with an enhanced labour rate (letter no. DS (G) 20/2000/49 dated 12 September 2008) and the net estimated amount fixed was Rs. 347.75 crore (DPR Rohmoria 2009).

In summary, the banks of the Brahmaputra River around Rohmoria have been continuously eroded for the last 50 years. Though the situation has been extremely hazardous, most State plans to check bank erosion have never been implemented. Until the oil blockade began, no tangible steps were taken.

The 'Palliative' measures at Rohmoria

The CWC, New Delhi vide letter no. 59/Rohmoria/01/FM-II 328 dated 16 July 2009 suddenly decided to scrap most of the earlier recommendations, waking up late to the realisation that a project was already being carried out under the aegis of the Brahmaputra Board, to close the Ananta nallah further upstream. The State government was informed that if the diversion of the Lohit River through the Ananta *nallah* could be closed, then the additional pressure and the subsequent slumping of the riverbank near Rohmoria would also stop. So instead, they suggested some immediate 'palliative' measures costing about Rs. 60 crore for Rohmoria. These were as follows:

1 Choking up all the southern incoming channels towards Rohmoria.
2 A channel guiding programme by excavating strategic paths to train the channels to retreat towards the middle path.
3 Bank stabilisation to check the high degree of scouring or toe-cutting of the bankline due to the loose sandy soil below the top soil.

In a nutshell, the project intended to revert the Brahmaputra River morphology back to its old status of 1915. Additionally, the banks were proposed to be made erosion resistant.

A comparative study of the 'New proposals', 'Suggested changes', and 'Sanctioned proposals' within a span of four months (16 July 2009 to 13 October 2009) clearly shows the absence of scientific research, and the preponderance of an *ad hoc* approach among the decision makers.

In the last part of year 2010 (16–22 September), a delegation from Rohmoria went to New Delhi to submit a memorandum to the president of India, the Prime Minister and the Secretary, Water Resources Department. A geo-fabric technology based project (Rs. 52.36 crore) was ultimately initiated in the month of January 2011. Rohmoria-Gora-Khahoniya-Protirodhi-Mancha took up the singular task to supervise the progress of work and maintain vigilance to ensure the quality of construction materials. Though by June 2011, 75 per cent of the work was completed, the target of finishing the task before the onset of rainy season could not be accomplished due to mismanagement in supplies, lapses in execution, etc. One important thing is any incomplete mission always bears the risk of getting completely washed away causing huge loss in the tax payers' money. Continuous reminders from the High Court resulted ultimately into the completion of bank stabilisation measure by the end of 2012 along a 2.6 km stretch of the most severely eroded bankline. Stacks of geobags were piled up and covered by steel wire mesh. Those stretches where geobags appeared to fail, porcupine stacks were put to check the bankline collapse. The expenditure incurred was Rs. 64 crore; much higher than the original plan estimate and yet only 40 per cent of the job was done. The good news is, the bankline stabilisation measure worked. Till today, that is 15 August 2015, no further erosion has taken place. Fear among the people living by the side of the bankline has subsided considerably; however, they complain about the incomplete execution of the project and virtual scrapping of the original plan to stabilise a 6.5 km stretch of the bankline. By doing a good work, the State had the opportunity to earn confidence of the erosion affected population. However, its vote-bank oriented electoral politics and a very casual *ad-hoc* approach to deal with the water borne disasters seems to be guided by the sole objective to clear up the oil blockades. As a result, in spite of the apparent lull in the protests and *dharnas*, scepticism about the bank stabilisation project is not over. The recent field visit by the author and interviews with the local inhabitants didn't show any sign of further State initiative. Moreover, no measures were visible to monitor and maintain the stabilised bankline and at a number of places, the steel wire meshes were found to be missing. In quite a large number of instances in the past, the River took unpredictable turns and the river science is still at its infant stage of development to predict accurately the near future nature of bankline migration. 'Palliative' measures have no doubt achieved its goal and the State should have initiated serious research on understanding the fluvial dynamics in the valley while encouraging the agrarian masses, the highly ignored stake holders to participate in the basin management programs. The State has avoided so far conveniently this task of furthering participatory democracy.

What is so different about the Rohmoria struggle?

1 The seriousness and urgency with which the Rohmoria struggle raised certain fundamental political questions proved its

genuineness, which compelled others to join in to secure their local mass base.

2 From the very beginning, the composition of the participants in the movement was highly diverse. The presence of workers from tea gardens was increasingly being noticed. Students and youths of both sexes participated. Housewives from villages participated very enthusiastically, and did not hesitate to address the masses.

3 This movement has helped to broaden the vision of the people. From the previous practice of getting quick short time relief, people have understood that the basic problems are far more serious. Ecological and political issues cannot be and should never be separated from each other. Thus, in about more than 35 years, the struggle of the people of Rohmoria has passed through different phases of surges and occasional periods of lull, without exhausting itself. In fact, the growth and enrichment continued. In the last 15 years, Rohmoria has become synonymous with the protracted resistance struggle of the toiling masses.

Conclusion

One of the most important causes behind the recent social disturbances among the farming communities in Assam is rooted in the massive rate of riverbank erosion which opens up a Pandora's box with a series of connected issues which deserve thorough research. It is high time that the State understands that wrong-headed models of progress and social negligence over decades may boomerang at unexpected moments with unpredictable intensities, unless a sincere effort is made to comprehend the processes behind the core issues of such a place and the people therein. Such an understanding needs to be integrated in a systemic strategy to address the issues at stake.

Acknowledgements

The author is thankful to Ranjan Gogoi and Montu Kumar Dutta, leading activists associated for a long time with the Rohmoria struggle, for providing data, old leaflets and unpublished reports of concerned State government departments and copies of official letters. The author is also thankful to the Indian Institute of Technology, Kanpur for providing institutional support to conduct some of the technical aspects of this study. Sincere gratitude is expressed to the India Office Library and Records, London, UK, for providing the topographic map of the study area prepared during 1912–26 seasons.

References

Goswami, D. C. 1985. Brahmaputra River, Assam, India: Physiography, basin denudation and channel aggradation. *Water Resources Research*, 21(7): 959–978.

Kotoky, P. D. *et al.* 2005. Nature of bank erosion along the Brahmaputra River channel, Assam, India. *Current Science*, 88(4): 634–640.

Lahiri, S. K. 2008. 'Resistance Struggle by the Erosion Affected People of Rohmoria: A Different History in the Making'. In: Sarma, C. K. (ed.). *North East India History Association Souvenir*. 29th Annual Session. Kokrajhar: North East India History Association, pp. 324–334.

Lahiri, S. K. and J. Borgohain. 2011. Rohmoria's challenge: Natural disasters, popular protest and state apathy. *Economic and Political Weekly*, 46(2): 31–35.

Lahiri, S. K. and R. Sinha. 2012. Tectonic controls on the morphodynamics of the Brahmaputra River system in the upper Assam valley, India. *Geomorphology*, 169–170: 74–85.

Lahiri, S. K. and R. Sinha. 2014. Morphotectonic evolution of the Majuli Island in the Brahmaputra valley of Assam, India inferred from geomorphic and geophysical analysis. *Geomorphology*, 227: 101–111.

Lahiri, S. K. and R. Sinha. 2015. Application of Fast Fourier Transform (FFT) in fluvial dynamics – A case study from the upper Brahmaputra valley in Assam. *Current Science*, 108(1): 90–95.

Molnar, P. and P. Tapponnier. 1975. Cenozoic tectonics of Asia: Effects of continental collision. *Science*, 189(4201), Aug 8: 419–426.

Narula, P. L., S. K. Acharyya and J. Banerjee. 2000. Seismotectonic Atlas of India and its environs. Kolkata: Geological Survey of India, pp. 1–40.

Sarma, J. N. 2008. 'Bank Erosion of the Brahmaputra River around Rohmoria in Dibrugarh District, Assam'. In: Sarma, C. K. (ed.). *North East India History Association Souvenir*. 29th Annual Session. Kokrajhar: North East India History Association, pp. 318–323.

Sarma, J. N. and M. K. Phukan. 2006. Bank erosion and bankline migration of the river Brahmaputra in Assam, India, during the twentieth century. *Journal of Geological Society of India*, 68(6): 1023–1036.

Thorne, C. R., A. P. G. Russell and M. K. Alam. 1993. Planform pattern and channel evolution of the Brahmaputra River, Bangladesh. *Geological Society Special Publication*, 75(1): 257–276.

Unpublished Detailed Project Report. 2009. Emergent Protection Measures for Protection of Rohmoria in Dibrugarh District (Revised), 2009. Submitted by the Executive Engineer, Dibrugarh Water Resources Division (WRD). (Referred in the text as DPR Rohmoria, 2009.)

11 The *char* dwellers of Assam

Flowing river, floating people

Gorky Chakraborty

Flood induced conflicts are seldom linear and are manifested over different spatio-temporal scales. They affect the victims according to their geographical location, ingenuity in adaption, resource endowment, institutional set-up as well as the capacity to influence state and non-state actors. Vulnerability to flood induced conflicts is, therefore, a function of a set of various social, economic and political factors woven together which may have both immediate and long term implications. Identifying the probable causes of flood induced conflicts with a nuanced understanding is of utmost importance for their resolution. This case study tries to understand the roots of conflicts and their manifestation at different levels of a flood prone population group living in the mid-channel bars (locally known as *chars*) of the Brahmaputra River and its tributary, Beki, in Assam (Map 11.1).

Geomorphology of *char*

Originating in a Himalayan glacier, the Brahmaputra River traverses a long route of 2,880 km covering four different countries (Table 11.1). During its meandering journey through Assam, it is joined by a large number of big and small tributaries, which influence the flow pattern, sediment discharge and bed load of the river in more than one way. The fluvial regime of the river is also greatly influenced by the characteristic feature of the Brahmaputra channel to migrate towards south. Various factors such as the existence of braided channels,[1] the gradient[2] of the river, average discharge (Table 11.2), the large quantities of suspended particles, sediment load (Table 11.3) and bed load, the geology and the seismic instability of the region[3] along with various other related factors[4] facilitate the process of *char* formation.

When the suspended particles and bed load flow through the braided channels of the river during floods, it gives rise to almond shaped alluvial formations, locally known as *chars*. Since they are formed under

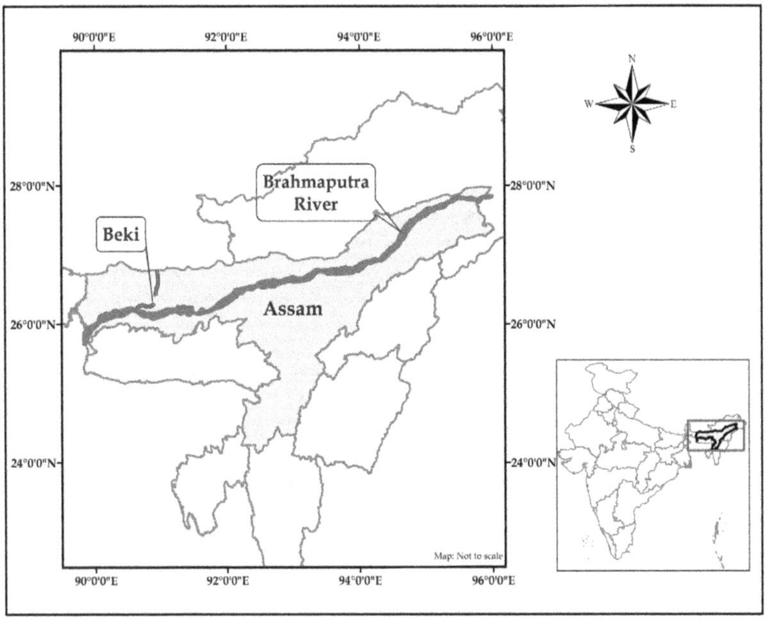

Map 11.1 Case study area

Source: SOPPECOM, Pune

Table 11.1 Area of the Brahmaputra Basin

Region	*Area (mha)*
China	**29.3**
Bhutan	**5.3**
India	**18.7**
Arunachal Pradesh	8.14
Assam	7.06
Meghalaya	1.17
Nagaland	1.08
West Bengal	1.26
Bangladesh	**4.7**
Total	**58**

Source: Singh *et al.* (2013: 92)

flood environment, the height of the *char* is never greater than the height of the highest flood. Once formed these unstable formations become an integral part of the fluvial system of the Brahmaputra, till they are eroded by next or subsequent floods. These geo-hydrological

Table 11.2 Average discharge of ten major rivers around the world

Sl. No	*River*	*Average discharge at mouth (10m)*
1	Amazon	99.15
2	Congo	39.66
3	Yangtze	21.80
4	Brahmaputra	19.83
5	Ganges	19.83
6	Yenisei	17.39
7	Mississippi	17.30
8	Orinoco	17.00
9	Lena	15.49
10	Parana	14.90

Source: Goswami (1998: 59)

Table 11.3 Sediment yield in ten major rivers of the world

Sl. No	*River*	*Sediment yield (tons/km/yr)*
1	Yellow	1,403
2	Brahmaputra (at Bahadurabad, Bangladesh)	1,128
3A	Brahmaputra (at Pandu, India)	804
3B	Irrawady	616
4	Yangtze	246
5	Mekong	214
6	Orinoco	212
7	Colorado	212
8	Missouri	159
9	Amazon	146
10	Indus	103

Source: Goswami (1998: 59)

formations follow a peculiar pattern of downstream migration as they are subjected to erosion on their upstream and deposition on the downstream. This affects the geometry and location of those *chars* (Bhagabati 2001).

Inhabitation of the *chars*

Historically, the *chars* of Brahmaputra were not inhabited permanently. There are few references (Powell-Baden 1892) of temporary habitations in limited number of *chars* mainly for floodplain cultivation of mustard during the winter months (*pam kheti*) in Upper Assam and as temporary shelters for livestock in Lower Assam, particularly in the undivided Goalpara District. It was only during the colonial period that revenue generation from unused land by means of agricultural spread became important and hence these places came under large scale habitation. The Census Report of 1911 for the first time mentioned the movement of the people from various densely populated districts of East Bengal (presently Bangladesh) to Assam (mainly Goalpara District) for bringing the wasteland under cultivation. It is estimated that till 1951 the total number of farm-settlers from East Bengal must have been around one to one and half million, which was between one-tenth to one-sixth of the then total population of the State (Goswami 1994). They were officially settled with more than 6,213,000 acres of land up to 1947–48 (Census of India 1951). The cropped area under cultivation increased from 2.40 million acres to 4.79 million acres and the area sown more than once from 0.21 million acres to 0.71 million acres between 1901–02 and 1947–48 (Guha 1977). This group from East Bengal had a major role to play in this transformation, and their biggest contribution has been large-scale cultivation of jute. There was not only increase in the acreage of the crops cultivated (Table 11.4) by them but also corresponding increase in the productivity of those crops (Figure 11.1) in the areas of their settlement (Doullah 2003). The State also gained revenue by exporting commercial crop[5] cultivated by these new settlers (Medhi 1978). Thus there was a complete overhaul of agriculture in the State, which would have been unthinkable without the advent of this group.

Table 11.4 Acreage under different crops in Assam, 1911–12 to 1947–48 (in thousand acres)

Crops	*1911–12*	*1947–48*
Rice	2,573	4,004
Cereals and pulses	109	255
Rape and mustard	238	310
Oilseeds	7	39
Sugarcane	30	60
Tobacco	8	20
Jute	58	210
Cotton*	35	34

Note: *crop not cultivated by migrant population

Source: Extracted from Chakraborty (2009: 57)

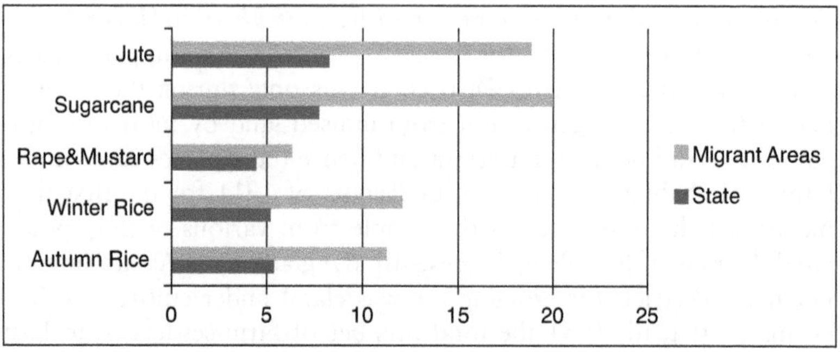

Figure 11.1 Percentage change in production per acre for selected crops in the four migrant prone districts vis-à-vis Assam, 1924–25 to 1947–48. Areas include four districts namely Goalpara, Kamrup, Nowgong and Darrang.

Contemporary socio-economic reality[6]

There is severe dearth of data concerning the *char* areas and their dwellers despite being inhabited for over a hundred years. There are two benchmark surveys (1992–93 and 2002–03) conducted by the Government of Assam, which are the only source of field level data of the *char* areas. If we compare both the reports it is evident that between the two surveys the number of *char* villages of the Brahmaputra has risen by 7.75 per cent but the population by 55.63 per cent (compared to 18.85 per cent for Assam between 1991 and 2001). The *char* dwellers comprise 9.37 per cent of the State's population but have only 4.6 per cent of the State's land and 4 per cent of its cultivable land. In fact, cultivable land as percentage of total land has declined from 70 per cent to 67.13 per cent during the time period. The density of population in the *char* areas (690 persons per sq. km) is more than double compared to that of the State (340 persons per sq. km). More than 81 per cent of the population is illiterate. Due to large scale flood and erosion the number of *char* villages in a district varies widely. For example, in 1992–93, Barpeta District had the highest number of *char* villages and population whereas during 2002–03, it was Dhubri. The poor socio-economic indicators and an extremely volatile fluvial regime where they inhabit is ultimately reflected in the poverty indicators of the *char* areas e.g. during 1992–93, 48.89 per cent of these dwellers were below the official poverty line which increased substantially to

67.89 per cent during 2002–03. Thus, migration in search of livelihood to the mainland areas remains the only option for their survival.

Flood, erosion and conflicts

A scan through the data provided by Assam Water Resources Department reveals that depending upon the nature of floods there are wide fluctuations in the average annual area flooded in the Brahmaputra Valley in the State. The reported average annual damage has been over Rs. 124 crore. The estimated area eroded since 1954 is 386,000 hectares, which have affected more than 90,700 families spread over 2,534 villages.[7] Estimated average annual erosion rate in the Brahmaputra Valley has been around 8,000 hectares. These State-level macro-data reveal a grim scenario but palpably much worse is the situation of those who are located within the river (the *chars*). However, there is no data regarding the plight of the *char* dwellers nor is there much discussion about flood-induced conflicts affecting their lives and livelihood patterns. In order to understand these scenarios, we have no other way but to depend on micro-studies.

A socio-economic survey (Chakraborty 2009) of 304 households in six *char* villages (four in Brahmaputra and two in Beki River, a tributary) spread over three development blocks of Barpeta District reveals that 29 per cent of the households were affected due to erosion and 48 per cent of the land was wiped out between 1988 and 2003. During this time period, the biggest flood occurred during 1988 and the damage due to erosion was also most severe as was its aftermath. More than 46 per cent of the total households affected by flood were victims of erosion. Forty-one per cent of the total eroded land so far was eroded during this year. During the next 10 years (from 1989 to 1998), when the floods were of lesser intensity compared to 1988, 45 per cent of the total households were affected and 51 per cent of the total eroded land was lost. Nearly 8 per cent of the total eroded households were affected and 7.60 per cent of the land was lost during 1999 and 2003. This shows not only the severity of erosion but also its recurring nature in these areas.

Another micro-study (Chakraborty 2006) of 22 households in Bechimari *char* village of Beki River in Barpeta District, covering the period of 1980–2004, reveals that more than 77 per cent of these households have become landless due to erosion and 94 per cent of their total land was lost. Members of the eroded households have migrated to mainland areas in search of livelihood options.

Once the households lose their land due to erosion, a spiral sets in whereby conflicts become a part of their daily existence with two broader ramifications – migration to mainland and claiming ownership to the eroded land if it re-appears after a while.

Migration of *char* dwellers and conflicts

When these victims of erosion migrate to the urban areas, they are often victims of labour market discrimination (Chakraborty 2011) which is reflected in two distinct forms. First, when the *char* dwellers migrate in search of livelihood to various districts of Upper Assam (where the people are less aware about the environmental reality in the *char* areas), they are subjected to forced eviction and deportation from these areas by pro-active groups in the name of liberating their homeland from the invasion of these illegal infiltrators (read *Bangladeshi*!). This amounts to denial in the labour market and results in exclusion by force and leads to conflict and violence.

Another phenomenon is observed in the urban centres of Northeast in general and Assam and Meghalaya in particular where there is seldom denial of entry of these victims of erosion as they are the source of cheap labour. But deprivation occurs as they are paid less than the prevalent wage rates for their manual labour (Dasgupta 2001–2002). Any protest on their part leads to job loss and conflict. It is often seen in both the examples of labour market discrimination that due to their sociocultural difference from the mainland population groups, they are branded as illegal immigrants (i.e. *Bangladeshis*) although the actual cause of their migration is flood and erosion occurring within the boundaries of Assam.

On the other, when these victims of erosion move to rural areas and natural habitats for livelihood options (mainly cultivation); it also gives rise to conflicts which are manifested over the long run. Often they settle in uninhabited land, which in the long run appears to be the *lived space* of the surrounding tribal population rooted in the practise of usufruct resource management. Conflict arises when the tribes claim these erstwhile unused lands and the *char* dwellers refuse to hand them over, which they have transformed into cultivable fields through their labour. The infamous massacre in Nellie (see Box 11.1) by members of Lalung and Tiwa tribes during 1983 and the recent incidents in the Bodo-dominated areas of Lower Assam are major indicators of this direction. More group-identity based conflicts over resource seem to be imminent under these conditions (Dasgupta 2001–2002).

Box 11.1 Genocide at Nellie

Nellie, otherwise a non-descript habitation, 40 km from Guwahati, is still remembered for the 18 February 1983 massacre during the heyday of the Assam Movement whereby 2,191 persons were butchered to death within a span of seven hours.

Although there are various interpretations as to what triggered this gruesome event (e.g. abduction of tribal girls, disregard to the election boycott call given by the supporters of the movement, etc.), where the victims were Muslim farm settlers historically hailing from erstwhile East Bengal and the alleged perpetrators belonged to the Tiwa tribe and other Assamese groups residing nearby. In hindsight it seems that land played an important role in generating palpable discontent, anger and frustration among both the communities over a long period of time, which was vented out on a different pretext on that fateful day.

Tribal land alienation has been an on-going process in this region. The 'colonial construct' of wastelands and their designs of facilitating the migration of peasants from East Bengal to cultivate cash crops and food crops in these wastelands on the one hand and on the other the problems of flood plains and erosion among these settlers and their migration to alternative sites among a host of other interdependent factors were destined to result in conflict. Apparently, the tribals' inclination to obtain ready cash by selling land at high prices and the willingness of the Muslim farm settlers to become landowners by arranging for it the sum/money seem nicely matched. Suruj Konwar, a veterinary department expert in Nellie says, 'When people get Rs. 30,000–40,000 per *bigha*, they simply sell their land'. Such processes continue unabated and grievances accumulate unknown and under the surface until one day, triggered by a petty cause, they cause a conflagration. Fellow citizens become mere *bodies* to vent their anger as the call beckons, *get rid of the Bangladeshis and save thy Motherland*. This is how Nellies are manufactured, and this is how Nellies are repeated!

Sources: Gokhlae, Nitin A. 2011. 'Who is responsible for Nellie massacre 1983?', 21 September 2011. www.ibtl.in/news/exclusive/1254/who.is-responsible-for-nellie-massacre-1983.

Hazarika, Sanjoy. 2000. *Rites of Passage*. New Delhi: Penguin.

Similarly, when they move into the forests or areas nearby, it leads to another set of conflicts. In these areas, often due to lack of proper documents they remain illegal settlers at the mercy of various State authorities, which itself becomes the cause for conflict generation. The *char* dwellers also become a threat to the surrounding ecology of the forest areas. Their haphazard settlements not only disturb the green cover but also become the source for human-animal conflicts in these areas. Studies suggest that the illegal settlement of these flood victims in the Doyang Forest Belt in Golaghat District of Central Assam has given rise to various environmental problems, the more severe being the human-elephant conflict, that have resulted in loss of life and property in and around the forest area.[8] The same study shows that in Burhachapori Wildlife Sanctuary and Kaziranga Forest areas the settlement of the flood victims has resulted in a rise in livestock population in these areas. Their presence has resulted in excessive grazing and loss of grassland and top soil cover. The domesticated animals have become the carriers of various diseases affecting the wildlife in the forest. The cumulative effect of such a situation leads to ecological marginalisation.[9]

Other than the urban, rural and forest areas, they occasionally also move to new *chars* and adjoining areas within the floodplain. When this type of migration occurs on a large scale or over a long period of time, it has a significant impact on the biodiversity of these natural habitats. The topographical maps of Survey of India prepared during 1911–12 show the presence of tall grasses (3–4 m in height) in various river islands of the Brahmaputra. Among these grasses there was also the growth of *Hemartheria pratesse* which is considered as the best fodder for the Indian one-horned rhinoceros and *Imperata cylindrical*, the main habitat of the endangered Bengal florican. With anthropogenic interventions these grasses have become extinct and the wildlife dependent on them has also ceased to exist in these areas. Again, senseless commercial exploitation of valuable species of trees such as *Lagerstroemia flos-reginae* (for boat making) has made them extinct in these areas. Certain weed varieties such as *Cyperus* (used mainly for mat making) are now under the threat of extinction because of over exploitation. Unmindful commercial fishing has led to depletion of marsh vegetation in these *char* areas.[10] Flood induced migration has thereby come into conflict with environment as well.

While the probability of erosion in the *char* areas remains high that of its re-emergence is also a natural, but an uncertain and unpredictable phenomenon. There are no legislative provisions to deal with such issues of re-emergence in Assam. This continues to be one of the main sources of conflict among the *char* dwellers. Neither The Bengal Alluvion and Dilluvion Regulation of 1825 (which was applicable in Assam for 32 years from 1897 to 1929) nor The Assam Land and Revenue Regulation 1886 deals with this issue directly. This legal ambiguity lays the ground for conflict. Due to the contentious nature of the problem associated with the re-emergence of land, there have been several court cases. Based on these disputes, certain outstanding judicial decisions[11] were promulgated, which to say the least, provided a way out for dealing with this issue of re-emergence.[12] It was argued that eroded land reformed on the old site continues to be the property of the owner, provided that it has not been completely abandoned meanwhile, where abatement of revenue is said to be an indication of complete abandonment. But Section 34(c) of the Assam Land and Revenue Regulation 1886 states that there is a provision for reduction of revenue by the Deputy Commissioner, for land, which has been washed away. But in reality, due to the attachment of large number of complicated explanations attached with this section, it is commonly believed that if an owner appeals for abatement of revenue for the part of his eroded land then there is maximum likelihood that he will lose the ownership of the entire eroded land.[13] Similarly, although there is provision, which states that, if the owner continues to pay revenue for the eroded land, after its re-appearance, the land will belong to the original owner. But there are conflicting references of time-period for which revenue has to be paid for an eroded land in various legislations. This inevitably leads to conflict among the contesting parties, where the scope of judiciary to deliver justice remains limited.

Another aspect related to land is observed among the *char* dwellers which on more than one occasion leads to conflict. In order to minimise flood related losses, the *char* dwellers have a natural urge to move from the peripheral regions of a *char* to the central area, which is the highest area within these alluvial formations. This takes place through land transaction, that is, sale of land in the peripheral areas and purchase in the central area. The same cycle is repeated when they migrate from an interior *char* to a comparatively less erosion prone *char* or settled areas within the flood plains. This risk minimising phenomenon of migrating from one place to another is a lifelong affair for a *char* dweller. A study[14] of the *char* villages of the Brahmaputra and Beki Rivers shows that out of the 304 households surveyed 24 per cent were involved in these kinds of land transactions during the one year period 2003–04. Among those involved in land purchase, 40 per cent shifted to safer places within their respective *chars*, equal number of households migrated to a comparatively stable *char* whereas the other 20 per cent moved to mainland areas. This migration of the *char* dwellers to newer areas particularly to the areas in the mainland often raises various questions about their identity and antecedents, which often is a source of conflict in those areas where they shift. Moreover, the sociocultural difference of the *char* dwellers vis-à-vis other population groups in the mainland (as observed in case of labour market discrimination) where they move in to settle, complicates the matter further.

Scope for dialogue and way forward

The first question that arises is dialogue between whom? Who are the contending parties? What is the scope and methodology for such a dialogue for conflict resolution? At the macro-level since there are no contending parties as such, the first task is to sensitise the non-*char* dwellers about the geomorphological details of the *char* areas and its consequences on the life and livelihood of the *char* dwellers due to their habitation in these alluvial formations. The state agencies, the media as well as the civil/community-based organisations should take the lead in this direction. The rationale behind the migratory nature of these people is basically flood-induced as they migrate from one location to another in search of livelihood options. Neither the *chars* are stable nor are its dwellers, so migration remains the only way of life for them. Migration also acts as an option to minimise risk and an instrument of insurance against their flood and erosion prone future. Until the non-*char* dwellers realise these ground conditions, it will be difficult to bring about a change in their perception of the *char* dwellers as intruders.

The location of these dwellers is geographically isolated and so they remain secluded from the psychological mainstream of the settled areas. Thus the issues related to the *char* dwellers are never a part of the dominant discourse. Under such a situation there is bound to be a sociocultural difference between the *char* and the non-*char* areas, however optimistic one may be about assimilation. The situation becomes more complicated as there is little presence of the State and its apparatus in these remote areas. The only proof of identity for the fortunate is the inclusion of their names in the electoral rolls. But during flood-induced migration they take along only the bare necessities of life, and a copy of the electoral roll as a proof of their identity is beyond imagination. The State has to intervene to allot some sort of identity proof to minimise conflict situation.

The issues related to land are of great significance for the *char* areas and the legislative gaps need to be filled up in order to minimise the chances of conflict. When the Goalpara Tenancy Act, 1929, Sylhet Tenancy Act (for Karimganj sub-division), 1936 and the Tenancy Act of 1935 were replaced by the Assam Temporarily Settled Areas Tenancy Act, 1971, it contained no reference to the *char* areas, which were at least sparsely mentioned in the earlier legislations. As a result, these areas remain outside the purview of land legislation. So when a new *char* emerges, it become the property of the State Government, which either keeps it as grazing reserve or settles it as per the Land Settlement Policy Resolution. But ground reality suggests that people in the densely populated *chars* seldom wait till its systematic settlement by the Government. They immediately occupy the newly formed *chars*. Power brokers play the deciding role, which ultimately leads to resource capture. Field level data[15] also substantiates the inequality of landownership in the *char* areas, the Gini co-efficient of inequality for landownership has been very high (0.60) among the surveyed villages. Without a proper legislative review for these areas backed by proper land settlement policies, conflict-resolution will be a far cry. Until measures are undertaken to normalise property rights, conflicts over issues of land will be rampant in these areas. The State Government needs to come out of its half-hearted effort for proper revenue settlements in the *char* areas (considering three failed attempts during 1979, 1983 and 1994) and start afresh by bringing in all the stakeholders together; else land issues will remain a recurrent source of conflict in the *char* areas.

Left to themselves, these flood-induced migrant *char* dwellers will float around in the region and beyond, and will be the perennial agents as well as victims of conflicts – both active and dormant.

Notes

1 A braided channel is one that is divided into several channels that successively meet and re-divide. Braiding in the case of the Brahmaputra in Assam seems to be influenced mainly by its high sediment load and weak band materials.

2 After flowing through the steep mountainous ranges in Arunachal Pradesh, the river gradient falls suddenly near Pasighat and then during its entire journey of 720 km through the plains of Assam it has a gentle gradient. The river has a gradient of 0.27 m/km at Pasighat, 0.09–0.17 m/km at Dibrugarh and 0.1 m/km near Guwahati. This makes the flow of the river slow which helps the process of char formation.

3 Major earthquakes of 1897 and 1950 have raised the river bed of the Brahmaputra, which inhibits the water flow and adds to the process of char formation. The bank line of the Brahmaputra is extremely unstable consisting mostly of fine sand and silt. Large-scale slumping of riverbanks

takes place when the water level falls after a flood, adding to the process of sediment formation. The river has a tendency for a lateral shift towards the south because of heavy silt discharge of the northern tributaries. The width of the Brahmaputra is therefore highly variable and can change dramatically from one year to another. This also adds to the process of char formation.

4 The unmindful construction of river embankments, along with construction related to railways and roadways, has disturbed the natural drainage in the state, which adds to the sediment generation in the river.

5 It is estimated that during 1947, out of the total quantity of exports from Assam, 62.5 per cent in case of railways and 37.5 per cent in case of waterways, comprised of jute, and overwhelming quantity of these jute exported were produced in the immigrant-prone areas, including the *char* areas.

6 This section is based on the two survey reports namely, Socio-Economic Survey Report, 1992–93 Char Areas Development Authority and Socio-Economic Survey Report, 2003–04, Directorate of Char Areas Development, Government of Assam.

7 These figures are based on the presentation titled 'A Review of Flood Management and Future Vision in Assam' made by Secretary, Water Recourse Department, Government of Assam on the occasion of 3rd North Eastern Council Meet, Guwahati, 9–11 March 2007.

8 Buragohain, P. P. 2005. 'Migrants and Their Socio-economic Status: A Case Study of Golaghat District', an unpublished seminar paper at a national seminar Tribal Demography of North East India, organised by UGC-SAP and Department of Economics, Dibrugarh University, 26 March 2005. For further details on human-elephant conflict in Golaghat district see, Talukdar, Bibhab, Kumar, and Barman, Rathin, 'Man-elephant conflict in Assam, India: is there any solution?' *Gajah*, Vol. 22, July 2003; Talukdar, B. K., J. K. Boruah and P. K. Sarma. 2006. Multi-dimensional Mitigation Initiatives to Human-Elephant Conflicts in Golaghat and adjoining areas of Karbi Anglong District, Assam Post Conference Publication of International Elephant Conservation and Research Symposium, Copenhagen, 21–22.

9 *Op. cit,* no.1.

10 *Ibid.*

11 There are few instances of legal disputes and promulgation of justice by the courts on issues related to land in *char* areas. Some of the cases include Lopez vs. Madan Mohan (13 MIA 467), Ramnath Tagore vs. Chandra Narayan Chowdhury (1 March 136), Imam Bandi vs Hargobinda (4 MIA 403), Radha Prasad Singh vs Ram Coomar Singh (3 Cal. 796) etc. Details are available for some of the cases. In the Boroji Manipurini (Appellant) vs. The State of Assam and Others (Respondent), 1958 (All India Reporter (AIR) 1958: 34) it was held that a land which has gradually and imperceptibly come out of the river-bed and added to the land of a riparian owner becomes part of the land belonged to him and is considered as his property . . . if it is considered an increment to the tenure of the land to which it has accreted. Similarly, in another legal dispute between Sudhangshu Ranjan Dasgupta ++++(Appellant) vs. Manindra Kumar Paul and Others (Respondent), 1972 (AIR 1972: 3 & 34), it was pronounced

that . . . any land accreted to plaintiff's periodic *patta* land by recession of river contiguous to such land automatically becomes part of the plaintiffs' land and cannot be allotted to any other person on the ground that the plaintiff had land in excess of ceiling fixed by The Assam Fixation of Ceiling on Land Holdings Act, 1957. For detailed analysis, see, Das, J. N. 1982. *A Study of the Land Systems of North Eastern Region.* Vol. I. Assam. Gauhati: Law Research Institute. Guwahati: Law Research Institute/Gauhati High Court, pp. 3–32.

12 Das, J. N. 1982. A Study of the Land Systems of North Eastern Region. Vol. I. Assam. Gauhati: Law Research Institute/Gauhati High Court.

13 *Ibid.*

14 *Op. cit,* no. 12.

15 *Ibid.*

References

Bhagabati, A. K. 2001. Biodiversity and associated problems in the Islands of the Brahmaputra, Assam. *Geographical Review of India*, 63(4): 330–343.

Buragohain, P. 2005. *Migrant and Their Socio-economic Status: A Case Study of Golaghat District.* UGC-SAP National Seminar, Dibrugarh University, 26 March, Dibrugarh.

Census of India. 1951. 3. Assam. Part I. New Delhi: Office of the Registrar General & Census Commissioner, pp. 71–72.

Chakraborty, G. 2006. 'Roots of Urban Poverty: A Note on Transformation of Entitlement Relationship'. In: Datta, R. B. and G. Das (eds.). *Informality and Poverty: Urban Landscape of India's North East.* New Delhi: Akansha Publishing House, pp. 180–192.

Chakraborty, G. 2009. *Assam's Hinterland: Society and Economy in the Char Areas.* New Delhi: Akansha Publishing House.

Chakraborty, G. 2011. From isolation to desolation: Analysing social exclusion among the *Char* dwellers of Assam. *Man and Society*, VIII, Summer 2011: 47–65.

Das, J. N. 1982. *A Study of the Land Systems of North Eastern Region.* Vol. I. Assam. Gauhati: Law Research Institute/Gauhati High Court.

Dasgupta, A. 2001–2002. 'Char' red for a lifetime: internal displacement in Assam plains in India. *South Asia Refugee Watch*, 3 and 4: 1–13.

Doullah, S. M. 2003. *Immigration of East Bengal Farm Settlers and Agriculture Department of the Assam Valley: 1901–1947.* New Delhi: Institute of Objective Studies.

Goswami, D. C. 1998. 'Fluvial Regime and Flood Hydrology of the Brahmaputra River, Assam.' In: Kale, Vishwas, S. (ed.). *Flood Studies in India: Memoir: Volume 41.* Bangalore: Geological Society of India, pp. 53–76.

Goswami, P. C. 1994. *The Economic Development of Assam.* Ludhiana: Kalyani Publishers.

Guha, A. 1977. *Planter-Raj to Swaraj: Freedom Struggle and Electoral Politics in Assam, 1826–1947.* New Delhi: ICHR.

Medhi, S. B. 1978. *Transport System and Economic Development in Assam*. Guwahati: Publication Board, Assam.

Powell-Baden, B. H. 1892. *The Land Systems of British India*. Vol. III. Oxford: Clarendon Press.

Singh, V. N. Sharma, C. Ojha, and P. Shekhar (eds.). 2013. *Brahmaputra Basin Water Resources*. Berlin/Heidelberg: Springer Science and Business Media, p. 92.

Talukdar, B. K. and R. Barman. 2003. Man-elephant conflict in Assam: is there any solution? *Gajah*, July: 50–56.

Talukdar, B. K., J. K. Boruah and P. K. Sarma. 2006. *Multi-dimensional Mitigation Initiatives to Human-Elephant Conflicts in Golaghat and adjoining areas of Karbi Anglong District, Assam*. Post Conference Publication of International Elephant Conservation and Research Symposium, 21–22 October, Copenhagen.

12 Seismic survey for oil exploration in the Brahmaputra River Basin, Assam

Scientific understanding and people's perceptions

Sanchita Boruah and Partha Ganguli

The Brahmaputra River, one of the largest rivers in the Indian subcontinent, is a virtual lifeline for the people of Assam. The Brahmaputra River Basin (BRB) is a rich reservoir of natural resources, including biodiversity and sedimentary deposits. The people residing on the banks of the Brahmaputra are dependent on the river either directly or indirectly for their livelihood. The river is a boon as well as bane; every year it brings floods mainly due to heavy rainfall and bank erosion due to changes in its braided course, which causes immense adversity and devastation. The proposed exploration in the river basin by Oil India Limited (OIL), a 50 per cent joint venture between the Government of India and Burmah Oil Company (BOC), may in fact add to the woes of the people, as it may disturb the flow of the river, and thereby affect biodiversity and livelihoods due to reduction in fish diversity, irrigation potential and agricultural productivity, and cause oil spills. A stretch of 170 km of the Brahmaputra riverbed (between Sadiya and Nimatighat) in upper Assam has been allocated to the OIL for conducting a seismic survey and later drill for oil.

The part of the river to be explored by OIL is at present being used by the people as a common property resource. Interference by OIL through a seismic survey in that part of the river will reduce the extent of the common property resource. Usually, interventions may disrupt people's occupations or provide alternative sources of income. Thus, they may ruin or augment their economy. However, the people feel threatened because there is a good possibility that they will not be compensated for their losses with better livelihood options. This chapter is an effort to objectively analyse and synthesise the conflict arising out of the seismic survey proposed by OIL, and people's perception and stand against it.

The need for exploration

Energy is indispensable for modern life. The demand for energy is multiplying with the passage of time and the growth of population. The search for energy has expanded its base from land to water and air to sunshine. The trend of energy consumption in India indicates the rising fuel consumption.

The largest oil reservoirs are called 'Super Giants', many of which were discovered in the Middle East. Because of their size and other characteristics, Super Giant reservoirs are generally the easiest to find, the most economic to develop and the longest-lived. The last Super Giant oil reservoirs discovered worldwide were found in 1967 and 1968. Since then, smaller reservoirs of varying sizes have been discovered in what are called 'oil prone' locations worldwide – oil is not found everywhere. Geologists understand that oil is a finite resource in the earth's crust, and at some future date, world oil production will reach a maximum – a peak – after which production will decline. This logic follows from the well-established fact that the output of individual oil reservoirs rises after discovery, reaches a peak and declines thereafter.

With close to 70 per cent of its oil requirements imported from more than eight countries, India is a net importer of oil. The rest 30 per cent is provided by the domestic oil production. India's oil consumption has increased but the production has almost remained the same with no discovery of Super Giant oil reservoirs.

Oil obviously provides significant benefits to society. Oil serves a wide diversity of purposes, which include transportation, heating, electricity and industrial applications, and is an input into over 2,000 end products. Oil is a high energy density abundant fuel, which is relatively easy to transport and store, and is extremely versatile in its end uses. The oil industry is phenomenally profitable for some corporations and governments. Taxes from oil are a major source of income for many governments. The global oil industry also provides significant jobs, profits and taxes.

Thus Northeastern region and especially Assam is constantly being explored for oil either to meet the energy requirement of the country or to save foreign exchange.

Chronology of historical events of OIL exploration in Assam

- Drilling for oil first began in 1886, by the Assam Railways and Trading Company, but it was not until 1889 that the first commercially viable well was struck at Digboi. India's petroleum industry thus started taking shape with the commercially successful discovery of oil in Digboi. The Assam Oil Company (AOC) was formed

in 1889 to take over the oil interest of the Assam Railways and Trading Company and the Assam Oil Syndicate, which had carried out the early drilling in the area.
- When oil was discovered in remote Digboi, there was no habitation in its immediate neighbourhood. The jungles were dark and swampy. The forests were so thick and the undergrowth so dense that sunlight could never reach the ground. Once oil was found, the dense jungles made way for the growth of the oil industry in Assam. Private oil companies mainly carried out exploration of the hydrocarbon resources of Assam.
- In 1953, the first oil discovery of independent India in Assam was made at Nahorkatiya near Digboi and then at Moran in 1956. The AOC was producing oil at Digboi (discovered in 1889) and was engaged in developing the two newly discovered large fields in Naharkatiya and Moran. The success at Nahorkatiya was the culmination of a long story of failure, frustration and despair in the oil exploration activities in Upper Assam. This triggered the initiation of oil exploration programmes elsewhere in the country.
- In 1955, the Government of India decided to develop the oil and natural gas resources in various regions of the country as part of public sector development. With this objective, an Oil and Natural Gas Corporation (ONGC) was set up towards the end of 1955, as a subordinate office under the then Ministry of Natural Resources and Scientific Research.
- Oil India Private Ltd. was incorporated on 18 February 1959 for the purpose of the development and production of the discovered prospects of Nahorkatiya and Moran, and to increase the pace of exploration in the Northeast. By a subsequent agreement on 27 July 1961, Government of India and BOC transformed OIL to a joint venture company (JVC) with equal partnership.
- Over the past 100 years, the petroleum industry of India with Digboi as its nucleus has helped to shape industrial development in the State of Assam.

Due to decrease in oil reserves worldwide and subsequent increase in the demand for oil, OIL then forayed into an exploration initiative in the Brahmaputra River Basin in India.

Methodology

Increased environmental concern and strict legal regulation adopted in oil exploration around the world has opened new avenues for many researchers to work on the possible effects of oil and gas exploration on different dimensions relating to environment and human societies. As a result, different groups adopt different approaches in order to assess the impact of such activity on the environment and humanity at large. In this chapter, our approach is based on compilation of the expressions and reactions of different groups in different forums and also individual opinions put forward by them.

Study area

The present chapter focuses on two particular districts of upper Assam which have the richest oil reserves, that is, Dibrugarh and Tinsukia districts (Map 12.1). These districts cover the area proposed for seismic survey by Oil India Limited in 2006.

Possible technology of the proposed seismic survey procedures to be used in the BRB

In the petroleum industry, the most common and effective method of obtaining a clear picture of the Earth's subsurface is seismic exploration.

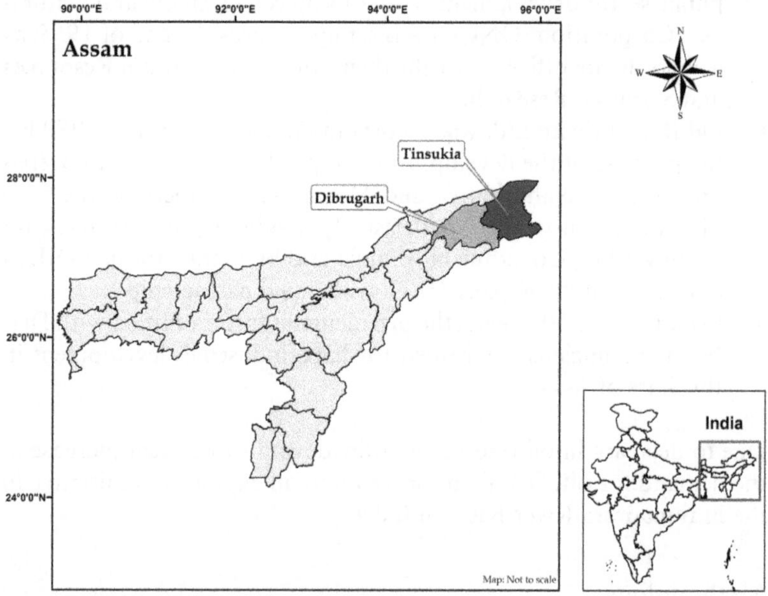

Map 12.1 Case study area: Dibrugarh and Tinsukia districts in Assam

Source: SOPPECOM, Pune

Seismic exploration is the use of seismic energy to probe beneath the surface of the earth, usually as an aid in searching for economic deposits of oil, gas or minerals. Seismic exploration, broadly speaking, creates a picture of the subsurface by recording vibrations as they bounce back from geologic formations. The resulting data is called seismic lines. Dynamite exploded underground is most commonly used to initiate the seismic waves. Upon arrival at the detectors, the amplitude and timing of waves are recorded to generate a seismogram (record of ground vibrations).

Generally, the density of rocks near the surface of the Earth increases with depth. Seismic waves initiated at the shot point may reach the receiving point by reflection, refraction or both. The results of a seismic survey may be presented in the form of a cross-sectional drawing of the subsurface structures as if cut by a plane through the shot point, the detector and the centre of the Earth. Such drawings are called seismic profiles. Dynamite explosions are used to create the seismic waves, and geophones laid out in lines measure how long it takes the waves to leave the seismic source, reflect off a rock boundary and return to the geophone. The resulting two-dimensional image, which is called a seismic line, is essentially a cross-sectional view of the earth oriented parallel to the line of geophones.

When seismic waves generated by dynamite reach a bedding plane separating rocks of different acoustic density, then a portion of the waves reflects back to the surface, causing the ground surface to rise or fall depending on whether the expansion or compression phase of the wave is being recorded. The remaining portion of the waves is refracted and diffracted. 'Wiggle traces' in a seismic line record the 'two-way' travel time (from source of generation to bed to geophone arrival time) of a reflected wave.

Why the conflict?

At present the oil industry of India is emphasising more on domestic oil exploration, increasing access to overseas oil and developing more refining capacity in the country. Although the case for the economic and political benefits of increased production and control over oil has been clearly articulated, the environmental, health and social costs of increased oil flows are largely absent from government policy deliberations. And perhaps more importantly, the equitable distribution of costs and benefits of increased oil production among communities and individuals is almost completely absent from public discourse. The proposed oil exploration on the BRB by the OIL has therefore come under attack from several organisations. The Asom Jatiyatabadi Yuba Chatra Parishad, Sodau Asom Mottock Yuba Chatra Sanmilan and the Toilokhetro Suraksha Sangram Samity have urged the company to take into consideration the impact it would have on the environment while conducting the investigation (*Assam Tribune* 2 November 2006).

Box 12.1 People's Perception on Seismic Survey on Brahmaputra River Basin

'Our survival is at stake and we will not allow OIL to carry out the survey even if it means death', Chandra Kamal Dutta, a school teacher, said. He lives in Rohmoria, 500 km east of Assam's main city Guwahati, which comes under the OIL survey and where most people grow paddy or catch fish from the Brahmaputra River to survive. The six-month survey, expected to begin this month, will involve underwater explosions and firing airgun shots into the riverbed to obtain seismic data. 'The loud bursts disturb, injure and kill aquatic life, as has been recorded during such surveys in other parts of the world,' Bibhab Kumar Talukdar, a local conservationist, said. 'We anticipate similar effects in the Brahmaputra that cannot be allowed.' The river, home to the endangered Gangetic dolphin and other marine life, originates in Tibet, and flows through Assam and Bangladesh before ending its journey in the Bay of Bengal. Residents of the area said the seismic survey would accelerate soil erosion along the river bank. 'Once oil exploration takes place in the river, there will be more pollution, dislocation of people living on the banks, and destruction of paddy fields,' Talukdar said.

The Brahmaputra has eroded huge chunks of land along its 750-km stretch. Experts say that vital parts of the farmland would go underwater eventually. Officials of Oil India, which supplies crude oil and natural gas to four oil refineries in the east, refused to comment on the exploration controversy.

(Protests in Assam over hunt for oil in Brahmaputra, 15 December 2006, 17:16 IST, Agency: Reuters)

Clearly, there are very real trade-offs resulting from increased oil production and consumption. But how well do policy makers and the public understand the costs and benefits of such a commitment to OIL? What data are available to evaluate the impacts of oil production and consumption at different stages in the oil life cycle? What evidence and analysis are available to compare trade-offs in security, economic development benefits, energy dependence, environmental harm, health costs and cultural consequences of increased oil production?

These questions never occupied prominence in the study reports before the survey came up for discussion.

The conflict context

This particular study conforms to the Modern Conflict theory as postulated by C. Wright Mills (Mills and Mills 2000), the founder of modern conflict theory. According to this theory, social structures are created through conflict between people with differing interests and resources. Individuals and resources, in turn, are influenced by these structures and by the 'unequal distribution of power and resources in the society'.

'Black gold' often brings hardship and misery to the societies where it is found. Petroleum-producing countries are plagued by corrupt and authoritarian governments, lopsided and unsustainable economic development, and violent conflicts. Foreign powers and their huge multinational oil companies often manoeuvre for control of the oil fields through clandestine operations or outright military intervention. In addition, disaffected rebels challenge governments hoping to win a share of the lucrative oil revenues. Environmental damage by oil extraction can spark protest movements, which are frequently met by violent repression. Boundary disputes between States over oil reserves also lead to violence. As worldwide oil and gas production peaks and consumer demand continues to rise, prices soar, increasing the likelihood of conflicts for this scarce resource. If the price of oil increases further, will it be possible for civil society to restrain national and multinational companies from exploring the Brahmaputra Basin? Shouldn't society adopt an informed, pragmatic and holistic approach towards it? These critical questions must be explored.

Emergence of the conflict over seismic survey in Assam

Initially, there was limited objection by the people to the exploration of oil on land. Monetary gain from compensation diffused possible protests, and minor conflicts centred around the payment of compensation. People had limited knowledge about the importance of ecosystem services and resources. At that time, OIL did not require any legal permission from the government. Since no field data was

Box 12.2 Activists' Reactions Relating to Seismic Survey on Brahmaputra River Basin

Guwahati, 15 June

The Krishak Mukti Sangram Samity (KMSS) on Sunday warned of strong resistance if the Oil India Limited (OIL) went ahead with its proposed seismic survey on the Brahmaputra riverbed after the verdict of the Gauhati High Court. Reacting to the Gauhati High Court judgment giving the nod to OIL to carry out seismic survey on the Brahmaputra riverbed, the KMSS which had hit the headlines by unearthing several anomalies in the Public Distribution System (PDS) under the Right to Information Act, said that the justice delivery mechanism had often failed to feel the sentiments of the toiling masses.

Source: *Assam Tribune*, 15 June 2008

available, they commissioned a team from Gauhati University (GU) in the year 2006 to undertake a study on the ecological aspects and possible impacts. Based on this EIA study submitted by GU in August 2006, OIL obtained a no objection certificate (NOC) from the Assam government and the Pollution Control Board of Assam (PCBA).

The GU study report was criticised by several environmental NGOs of the State and declared as biased and misleading with incorrect facts and conclusions. The controversial report led to the first manifestation of conflicts with the NGOs over the issue of seismic survey in the Brahmaputra River. A public hearing was organised by OIL on behalf of the PCBA to acquire an NOC from the Ministry of Environment and Forests (MoEF). The public hearing was suspended without a solution. The proceedings did not end on a conclusive note due to protests and slogans from the people who demanded greater transparency in the environmental impact assessment (EIA) process. It was, however, decided that all queries will be handed over to officials of the PCBA. OIL was asked to forward their views on the public hearing and clarify the queries raised by the house. It was decided that the PCBA will again call for a public opinion on the same and until such time the survey work will remain suspended.

Later, A Multi-disciplinary Advisory Group (MDAG) was formed in June 2007 to re-assess the EIA prepared by GU. Research proposals from various expert groups were sought. The present status is that the license of the OIL to carry out the seismic exploration has expired, and is awaiting renewal.

Box 12.3 WWF-India/NGO's Concerns Regarding the Environment Impact Assessment

WWF-India welcomes the decision to suspend seismic work in Brahmaputra after a recent public hearing where environmental organisations voiced their concerns. The public hearing on the much-debated environmental impact assessment (EIA) conducted by Gauhati University was organised by Oil India Limited (OIL) which wants to drill Brahmaputra for oil. Long deliberations were witnessed that highlighted gaps found in the Executive Summary of the EIA report which was circulated for the general public.

The main concerns with the EIA report are:

1 The complete EIA was not made public.
2 The executive summary provided to the public for discussion in the meeting had a lot of incomplete information and was also found to be self-contradictory.
3 Impact on the social and economic structure of the human population to be influenced is not holistically analysed.
4 Impact on the ecosystem in general and the endangered species like the Ganges River Dolphin and other aquatic species is not properly analysed.
5 Impact of the seismic waves on the erosion and flood problem of the Brahmaputra River system is also not addressed.

WWF demands that a detailed and fresh EIA study is conducted as the region is extremely sensitive and very rich in biodiversity with participation of all stakeholders including conservationists and cetacean experts.

Source: Press release by WWF-India, 3 December 2006

Conclusion

Oil exploration and production creates significant and varied negative impacts on and costs to human health, cultures, and the environment. Thus, it is critical to evaluate the costs as well as the benefits of oil. Although government policies encourage more oil development, it provides little information on the negative consequences of this development. Instead, the reports cite only technological advances that have minimised the impacts of oil exploration and refining. Oil as a whole provides significant benefits to the society as it serves a wide diversity of purposes and provides direct and indirect employment to many. The north-eastern part of India is marked by underdevelopment and unemployment – one of the prime reasons for youth unrest and dissatisfaction of the society. The river being the lifeline for most of the people of Assam and a symbol of culture and tradition, the issue of conflict of interest was triggered. The employment generation through oil exploration is welcomed by a majority of the river people, but apprehensions persist about the ownership of 'common property resources', in this case, the Brahmaputra River and its basin. At this moment, we may conclude that OIL's lack of transparency and a hurriedly prepared dubious EIA report seem to be responsible for the conflict. At the end of 2011, the situation seems to be in a state of suspension. Even though the turmoil has subsided, once the oil companies resume explorations in the BRB, there is every possible chance of the re-emergence of the conflict.

Reference

Mills, K. and P. Mills. 2000. *C. Wright Mills: Letters and Autobiographical Writings*. Berkeley: University of California Press.

13 Hydrocarbon extraction in Manipur and its impact on Barak downstream

Jinine Laishramcha

Jubilant Energy Private Limited was awarded two blocks of oil deposition namely Block AA-ONN-2009/1 and Block AA-ONN-2009/2 that covers about 4,000 sq. km in the southwest of Manipur after the eight round of bidding under the New Exploration License Policy by the Government of India. The total area includes 2,217 sq. km of Block 1 in Churachandpur District and 1,740 sq. km of Block 2 in Tamenglong District and Jiribam sub-division which fall under the Assam-Arakan Basin. The production share contracts were signed by S. K. Srivastava, Director General of Hydrocarbons, and D. N. Narasimha Raju, Joint Secretary on behalf of the minister of Petroleum and Natural Gas, Government of India in July 2010. The Manipur Government granted the Petroleum Exploration License to Jubilant in September 2010. And recently civil society found another Block of about 220 sq. km in Jiribam subdivision of western Manipur given contract lease to Oil and Natural Gas Corporation (ONGC), India.

Alfa Geo Company and other sub partners have conducted seismic surveillance and other initial work of the project. In the environmental impact assessment (EIA) reports of Jubilant Energy about 30 oil wells have been identified in the mountainous terrain of Manipur where Barak River originates.

This impending hydrocarbon project is at the source of Barak River (Map 13.1 for the location of Barak River in Manipur), the river which is a principal source of livelihood and environment in the Indo-Bangla region. It will be a logical apprehension that the project will cause deterioration in the river system and environment, and will shrink the arable lands and biological resources. The condition will lead to a fierce contest for resources among the ethnic and religious communities in the region. Eventually the escalation of violent conflict is very proximate.

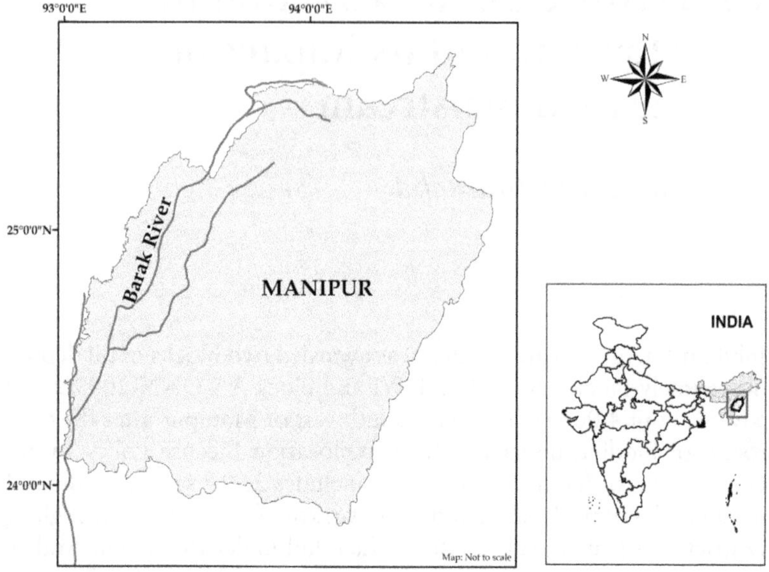

Map 13.1 Location of the case study area
Source: SOPPECOM, Pune

Protests

Civil societies, student organisations in Manipur and village representatives of likely affected area have been expressing their concerns about oil exploration and drilling activities since 2011. There have been a series of protests against the new project to discontinue the activities of Jubilant Energy and its sub-contractors (Figure 13.1).

The first meeting of civil society workers of different communities of the indigenous peoples of Manipur was held on 8 October 2011 in the conference room of the Manipur Film Development Corporation (MFDC), Imphal. In 2012, the following meetings and protests have taken place:

- On 14 March, a public meeting and rally were organised at Nungba Bazar in protest against oil exploration and future extraction. The local population, civil societies, human rights activists, students and community leaders participated in the protest.
- On 16 July, various civil society groups and other organisations of different indigenous communities organised the State Level Convention on Petroleum Exploration at the Manipur Press Club, Imphal. Important personalities and experts across Manipur

Figure 13.1 People's protest at Nungba public hearing

Source: www.epao.net

presented their critical analysis of the issue. The convention was attended by representatives from the Zeliangrong Students' Union, Zeliangrong Baudi, Naga Women's Union, North East Dialogue Forum, Zomi Human Rights Foundation, Singlung Indigenous Peoples Human Rights Organisation, Action Committee against Tipaimukh Project, Citizens Concern for Dams and Development, Tamenglong VA, Naga People's Movement for Human Rights and others. The convention resolved that the project should be stopped.

- On 29 July, a public meeting comprising of civil societies and locals at the Jiri Ima Meira Paibi Apunba Lup (JIMPAL) office in Jiribam reaffirmed that the exploration activities should be stopped and that the public hearing should not be allowed to be conducted by Manipur Pollution Control Board (MPCB) the following day there. However on 30 July, government agents and officials of Jubilant Energy were conducting the public hearing for Block 2, though public were not present.
- On 8 August, the Public Hearing for Block 1 was conducted at the Parbung Community Hall, Churachandpur District. At this hearing, the local people questioned the government and project officials. Following this event, Hmar, Zomi, Zelianrong students and human rights groups reacted strongly to the undertaking of the government and Jubilant Energy which discriminated against indigenous peoples and violated the UN principle of free prior informed consent.

- On 17 August, one of the scheduled public hearings for Block 2 at Nungba Town Hall, Tamenglong District was disrupted by locals and students.

Conflict linkages between Barak River and hydrocarbon project in Manipur

Hypothesis: the upcoming hydrocarbon extraction in Manipur will escalate the existing conflict in the Northeast India.

This extraction project in Manipur is a very critical concern for it will worsen the conflict dynamics among the communities in the Barak (Meghna) Basin right from its source in Manipur through Assam, way up to Bangladesh. The adverse impact of the water contamination due to the hydrocarbon extraction (Map 13.2) will make the livelihood

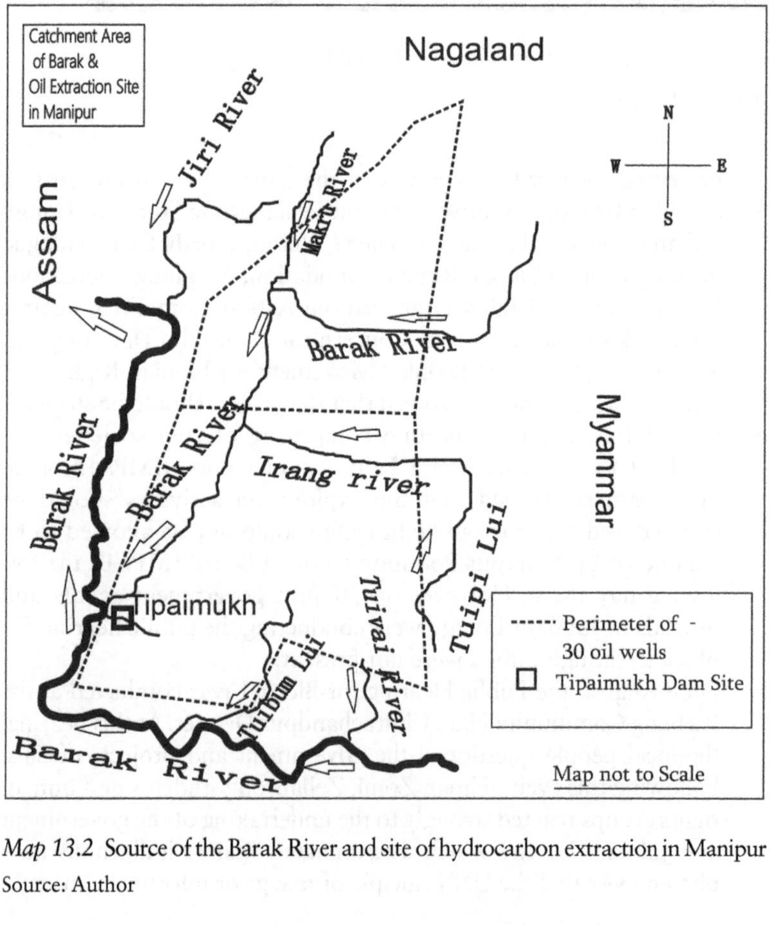

Map 13.2 Source of the Barak River and site of hydrocarbon extraction in Manipur

Source: Author

challenge more complicated in the region. The condition will intensify a fierce contest over the ever shrinking resources of land and water among various ethnic and religious communities (Figure 13.2 and Figure 13.3).

During the course of extraction, the saline (formation) water will come out; accidental oil spillage and frequent leakages will happen; sewage, surface runoff, drilling cuts and other hazardous oil contents will also be disposed. They will flow down through the creeks and drainages to the tributaries of Barak, namely Makru, Tuivai, Irang, TuipiLui and TuibumLui Rivers.

By taking all the pollutants from these small rivers into its stream, a stronger and bigger Barak will run down to the riparian of Surma

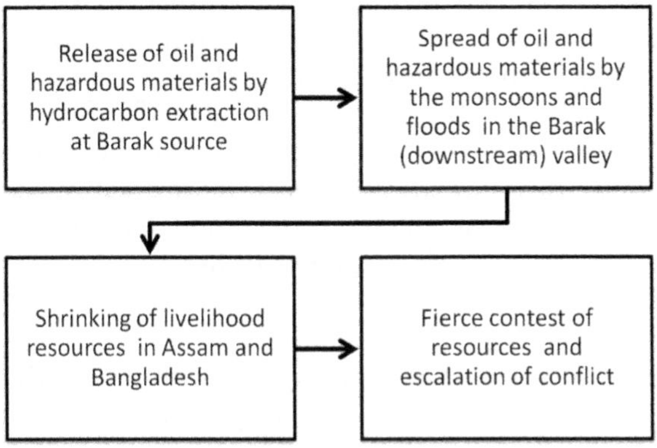

Figure 13.2 Proximate causes, conditions and eventuality of the hydrocarbon project in Manipur

Source: Author

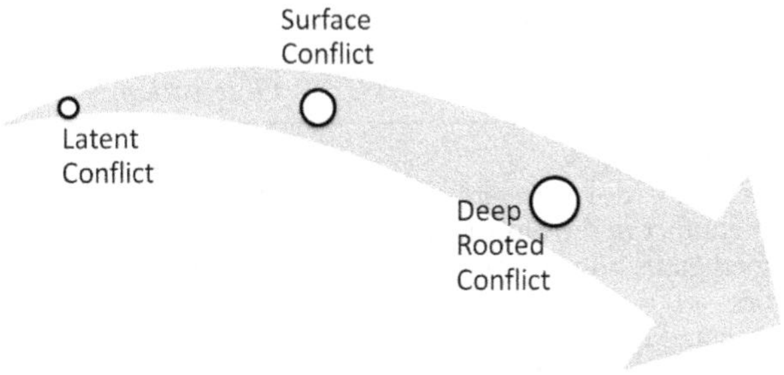

Figure 13.3 Shifting tendency of conflict in the downstream along the Barak River

and Kushiyara in Bangladesh after passing Assam. The pollutants will spread along the total length of Barak i.e. about 1000 km from its source in Manipur up to the mouth in the Bay of Bengal. The adverse environmental impact will spread relatively faster in the downstream because of two reasons – one, river flow will be buoyed up by the monsoons at the mountainous source of the Barak that recorded an average annual rainfall of 2000 mm, and second, frequent widespread floods in the Barak downstream.

In general, hydrocarbon production and transport has a significant impact on the landscape and local environment. Contamination of soil and water is a common consequence of oil production. In Ecuador, for example, oil and water separation stations in the Oriente generate more than 3.2 million gallons of liquid waste each day, most of which has been discharged untreated into the environment. Groundwater is particularly susceptible to contamination from the Formation Water, extracted along with oil during drilling. This is contaminated both with oil and heavy metals and is therefore toxic. Further impacts stem from the burn-off of excess natural gas which has had a devastating effect on water quality and biota there, most notably Nigeria. This burning releases methane, sulphur dioxide and toxic compounds. The dry winter can also exacerbate the risk of fires.

The stakes, sources and associating elements of conflict

Since this hydrocarbon project will damage the biological environment and consequently trigger harsh contest over livelihood resources among the communities; and the river systems, agricultural lands, wetlands, forests, floods, etc. will be the stakes, sources and associating elements of the looming conflict in the region (Table 13.1).

Table 13.1 Potential conditions contribute to the conflict

Potential factors for the conflict	*Observation*
Drainages at project sites	Hilly terrain and faster current
Flood in the Barak (downstream)	Frequent and widespread
Antecedent conflict in the Northeast India	Existing
Ethnic and religious hostility among communities	Existing
Militarisation and non-state armed groups	Existing

Source: Author

Barak River and its valley

The Barak flows southwest in Manipur and turns north in the Manipur-Mizoram border then flows into the Cachar District of Assam, where it moves westward near Lakhimpur as it enters the plains. The river goes through Silchar town, where it is joined by the Madhura River. After Barak traverses about 564 km in Manipur and Assam it enters Bangladesh.

The geographical area which is known as Barak Valley after the name of Barak River is the most southern part of the State of Assam. The Valley covers a geographical area of 6,922 sq. km of which some 3,839 sq. km (55.46 per cent) comprises forests that are mostly confined to the peripheral hill areas. The central plains abound in wetlands that occupy a total area of 13,737.5 hectares.

The valley comprises of three districts, namely – Cachar, Karimganj and Hailakandi. This is the meeting place of various ethnic and religious communities, such as – Khasi, Garo, Mizo or Lusai, Naga, Bishnupriya, Meetei, Hmar, Halam, Santal, Orang, Munda, Assamese, Bengalee, etc. There is both similarity and dissimilarity in socio-economic and cultural life of every community. All of them believe in their own ethnology and tradition.

The biological capital is the main strength of the valley. It includes the resources of agriculture, forest, river and wetland. Sylhet and Cachar were often dubbed as the granaries of Bengal and Assam, for their bountiful production of paddy. Fishing is one important economic activity in the region. Fisheries, especially in large floodplain wetlands/lakes (locally called *haors* or *beels*) like Chatla, Jabda and Lucca in Cachar, and Shonbeel and Ratabeel in Hailakandi and Karimganj districts, are major source of livelihood.

The vegetation is mostly tropical evergreen and there are large tracts of rainforests in the northern and southern-eastern parts of the valley, which are home to tiger, elephants, malayan sun bear, capped langur, etc. Rare species found are hoolock gibbon, phayre's leaf monkey, pig-tailed macaque, stump-tailed macaque, masked finfoot, white-winged wood duck, etc.

Flood

The flood in the Barak riparian will be one of the much potential driving forces to deteriorate the biological environment; the oil pollutants come in the Barak will dreadfully be spread in such a short period time of a single monsoon. The oil will reach wherever flood goes. In other words, the major part of the geographical plane bears the risk of suffering from oil pollution. The Barak Valley is commonly known as floodplain in southern Assam. During the monsoons the valley is usually flooded (there were three major floods during 1986, 1991 and 2004). The plight of the internally displaced people will be a mammoth challenge as the affected people will migrate to other areas and also move to the elevated land not affected by the oil pollutants.

In Bangladesh floods are more or less a recurring phenomenon. Each year in Bangladesh about 26,000 sq. km, 18 per cent of the country is flooded. During severe floods, the affected area may exceed 55 per cent of the total area of the country. About 40 per cent of country flood occurs in Barak downstream i.e. Surma, Kushiyara and Meghna. Flash flood occurs due to heavy rainfall in Barak Basins in India.

In addition to possible direct impact of the oil pollutants to biological environment, one secondary adverse impact will be on the migrant rice-worker from lower Meghna such as greater Comilla, Dhaka, Noakhali, Faridpur and even from Barisal area. They come to the *haor* (depression) areas of Surma and Kushiyara to work on harvesting and processing *Boro* (winter) rice. These workers work as exchange worker in which cash is not transacted; they get share of the rice harvested from the farmers.

In the Surma trough, the *haors* and the arable lands require full flood and drying of the land to maintain the ecological and hydrological balance. The flooding is generally considered removing of the undesirable and toxic chemicals left over by the farming, fish droppings and the natural wastes left over by the millions of migratory birds. In addition, the aquatic plants, peri phytons, bacteria, fungus and algae or other microbes require flushing every year. The flooding especially the early flood cleans the river beds and banks to a level so that the spawning fish can make their nest for the hatchlings. The flooding also provides the young fish ample space for play and movement including their uninterrupted travel during their migration downstream. However, due to the oil extraction in Manipur, the desirable contribution of flood will be changed completely into the uttermost ugly reality of human suffering and environmental destruction.

The riparian region of Barak in Assam

Most of the lands for human settlement and cultivation in the following three districts are consolidated by the Barak River.

1 Cachar: 3,786 sq. km, population about 1.8 million, population density 500 per sq. km.

2 Hailakandi: 1,326 sq. km, population about 0.7 million, population density about 500 per sq. km.
3 Karimganj: 1,809 sq. km, population about 1.3 million, population density about 700 per sq. km.

The Barak is a principal water source in the valley for various purposes of human and other biological system including agriculture, fishing and other activities. This intensely meandering river covers extensive plain area of the southern Assam.

The Barak passes important towns and villages, Silchar, Badarpur, Panchgram, Katakhal, Salchapra, Kalinagar, Srikona, Ramnagar, Masimpur, Sonai, Banskandi, Lakhipur, Udharbond, Jirighat, Borkhola, Katigorah, Dholai, Joypur, Kachudaram, Fulertal, etc. Population density is observed to be highest along the river. Agricultural activities and home gardens are characteristic along the entire stretch of the river.

Silchar, the headquarters of Cachar District, is a town of 15.75 sq. km with a population of around 0.2 million and population density of about 13,000 per sq. km. A major part of the district is encircled by the Barak. It is one of the busiest towns of Northeast India and a commercial hub for the States of Tripura, southern Assam, Manipur and Mizoram. Approximately 90 per cent of the residents of Silchar are Bengalis who speak Sylheti dialect, the rest being Dimasa Kachari (Barman), Meetei, Marwaris, Bishnupriya, Assamese and some to tribal communities like Nagas.

The riparian region of Surma and Kushiyara in Bangladesh

After Silchar, the Barak flows for about 30 km. Near Badarpur, the river divides itself into the Surma and the Kushiyara and enters Sylhet division in Bangladesh. The Meghna is formed inside Bangladesh above Bhairab Bazar, by the confluence of the Surma and Kushiyara. The Barak covers a total length of about 468 km in Bangladesh.

Between Surma and Kushiyara, there lies a complex basin area comprised of depressions (*haors*) usually used for fishing and fishery. Most of the Surma system falls in the *haor* basin, where the line of drainage is not clear or well defined. In the piedmont tract from Durgapur to Jaintiapur, the network of streams and channels overflows in the rainy season and creates vast sheets of water which connect the *haors* with the rivers.

Box 13.1 The course of River Barak

1 Sylhet District with an area of 3,490.40 sq. km is bounded by the Khasia-Jainta hills of India, population 3.5 million; Muslim 91.96%, Hindu 7.80%, Christian 0.09% and others 0.15%; ethnic nationals – Khasi (Khasia), Meetei and Patra (Pathar); main occupations - agriculture 30.82%, agricultural labourer 15.59%, fishing 3.6%; cultivable land 66%, fallow land 34%; single crop 54%, double crop 36% and treble crop 10%; main crops and fruits – paddy, mustered, betel nut, mango, jackfruit, orange, litchi. Fisheries, dairies, poultries – fishery 110, cattle farm 112, dairy 12, poultry 228, hatchery 8.

Sylhet City lies on the bank of the Surma with a population estimated at 0.5 million in an area 10.49 sq. km, density population about 50,000 per sq. km; 85% of the population of Sylhet is Muslim. Other religious groups include Hindus (15%), and Buddhists and Christians (less than 0.1%).

2 Sunamganj District with an area of 3,669.58 sq. km with many *haors* and *beels* is bounded by Khasia and Jaintia hills. Population about 2.5 million; Muslim 83.62%, Hindu 15.95%, others 0.43%; ethnic nationals 6,643 (Meetei, Khasia, Garo and Hajong); main occupations – agriculture 43.43%, fishing 3.34%, agricultural labourer 24.10%; arable land 294,021 hectares, fallow land 51,752 hectares; main crops and fruits – boro paddy mango and orange; fisheries, dairies, poultries – fishery 604, dairy 105, poultry 697, hatchery 6.

Sunamganj Town stands on the bank of the Surma. The town has an area of 22.16 sq. km It has a population of about 52 thousand; density of population is about 2,500 per sq. km.

3 Habiganj District with an area of 2,636.58 sq. km, is bounded by Tripura State of India on the south. Population 21 lakhs. Main occupations – agriculture 42.26%, agricultural labourer 20.55%, fishing 2.73%; total cultivable land 25,299.3 hectares, fallow land 520.53 hectares; single crop 51.6%, double crop 38.7% and treble crop land 9.7%; main crops and fruits – paddy, tea, wheat, potato, jute, ground nut, betel leaf, oil seed, mango, jackfruit, banana, litchi, coconut, lemon, pineapple and black berry; fisheries, dairies, poultries – (dairy 148, poultry 739 and fishery 638).

Dynamics of the conflict

At the source of the Barak

The State is already home to large number of insurgent groups fighting for various causes ranging from creation of homeland to protecting the identity of the ethnic communities. In this respect, the issues related to hydrocarbon extraction and dam building may further disturb the already fragile situation in Manipur. It might certainly add to the sense of disgruntlement within the ethnic communities and create fresh batch of insurgents. Moreover, the large presence of the Indian army and paramilitary forces with the impunity of the controversial Armed Forces (Special Powers) Act, 1958, and their contentious activities, will continue to destabilise normal life. The fear of the human rights violations by the security personnel and their alleged nexus in the inter-ethnic and inter community violence will prolong situation of the armed conflict in Manipur.

In the downstream

The impact of the project (Map 13.3) may be critical in the downstream because, first, the water will not be fit for human consumption, agriculture, the environment, flora and fauna. Second, the frequent floods of the Barak Valley and of Sylhet division will spread the oil and other hazardous pollutants through the riparian region. There will be long-term deterioration of land, water, vegetation, air and public health. One of the crucial impacts will be of the destruction of paddy fields and fishing areas. Rice and fish is the staple food in the region. Most importantly, water shortages will become acute. For instance, Sylhet City faces severe water scarcity. The city corporation is supplying only 22,500 gallons of water, far less than the requirement of about 65,000 gallons.

Economic activities will be ruined, people will lose their livelihoods. Consequently, conflicts over the limited livelihood resources will exacerbate affecting the population of 3.8 million of southern Assam of India and 9.8 million of Sylhet Division in Bangladesh.

The challenge may be abetted with the various aspects: the demographic pattern itself display that a potential for communal and religious strife is very high. The religious break-up of the population in Barak Valley; Hindu 42 per cent, Muslim 50 per cent, Christians 4 per cent and others 4 per cent. Hindus are majority in Cachar District (57) per cent and Hailakandi District (58 per cent) while Muslims are majority in Karimganj District (53 per cent). A great

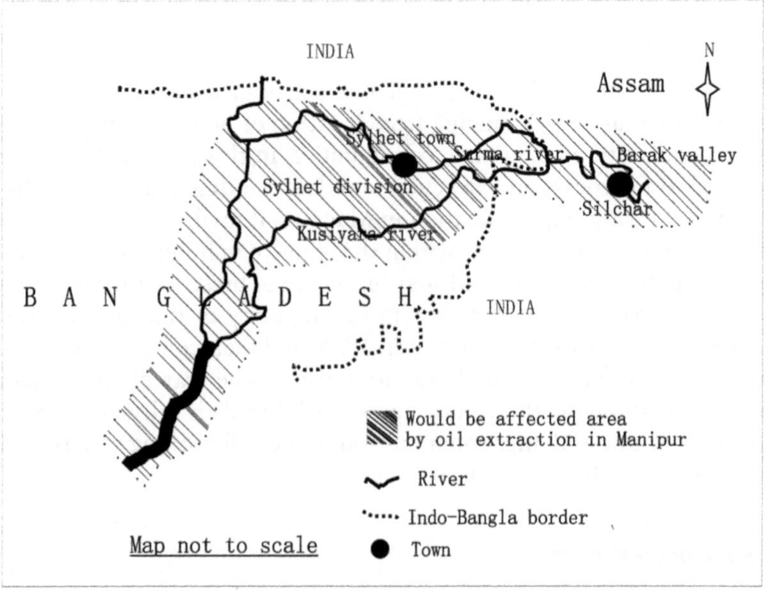

Map 13.3 The areas in the Assam and Bangladesh will be affected by the hydrocarbon project in Manipur

Source: Author

apprehension is that the violence may not be confined to certain geographical locations, but could spread across the entire Northeast India namely Assam, Arunachal Pradesh, Manipur, Mizoram, Meghalaya, Tripura, Nagaland and across Bangladesh. This oil project impact will add-up to another cause to the already existing conflict of the region.

The information revealed by the UN Department of Economic and Social Affairs (UN-DESA) shows that in 2013, India was home to 3.2 million Bangladeshi residents who had migrated into the country. Though there is no State-wise break-up of the Bangladeshi migrants, the problem is most severe in Assam.

Thus the emergence of a surface conflict may become more visible than ever before in such a very vulnerable atmosphere. The nature of the possible conflict and violence may be very similar to the communal violence in Kokrajhar, Chirang and Dhubri in Assam which is caused by land and livelihood issue rather than the religious hostility.

Here, we may acknowledge the condition and cause of the recurring Assam conflict. There is implication of 'outsiders' encroaching on 'others' land. Many unskilled or semi-skilled people have crossed over from Bangladesh to Assam and neighbouring States in search of livelihoods. It is a fact that for those living in Bangladesh's border areas with a population density of 1,150 per sq. km and a per capita income of Rs. 46,870, Assam with a population of 397 per sq. km and a per capita income of Rs. 84,400 is a greener pasture. According to the Supreme Court of India, the all India percentage of decadal increase in population during 1981–91 is 23.85 per cent, whereas in the border districts of Assam, the decadal increase is 42.08 per cent in Karimganj, 47.59 per cent in Cachar and 56.57 per cent in Dhubri. It can be assumed that the infiltration of foreigners from Bangladesh contributed significantly to the sharp increase in Assam's population.

Owing to the impact on the downstream by hydrocarbon extraction, the conflict in the region may turn more violent; the latent conflict have already turned into a surface conflict, which may in turn become deep rooted (Figure 13.3). The conflict could be intracommunity – within Hindu or Muslim or ethnic communities themselves; intercommunity – among the Muslims, Hindus, Christians and others; between the State government and their peoples; interstate – more or less among the different peoples of the Indian States (at the peoples level); among the Indian States (at the State governments level) and between the peoples of Northeast India and Central government, and international and trans boundary – between India and Bangladesh. An exacerbation of the armed conflict is very likely.

Conclusion

The campaign for sustainable development has been revitalised by the United Nations Conference on Sustainable Development, Rio+20, which took place in Rio de Janeiro, Brazil, in June 2012, 20 years after the landmark 1992 Earth Summit. It underlines that we are to ensure that the world is liveable for our children and grandchildren, with a greener, sustainable economy. Towards this end, we must consume energy wisely. Sustainability calls for a decent standard of living for everyone without environmental deterioration, and without compromising the needs of future generations.

The two EIA reports for the two blocks of hydrocarbon project in Manipur prepared by SENES Consultants India Private Ltd. for Jubilant Company mentioned possible damage to the biological environment, the flora and fauna of the forests, aquatic ecology, water resources, demography, and ambient air. A critical impact of oil extraction that is not mentioned fairly in the assessment reports is the long-term pollution of the surface and ground water resources due to drilling activities, the release of saline water, accidental crude oil spillage, etc. A very important potential pollutant known as Formation Water is not highlighted in the EIA reports. When crude oil is produced, this pollutant comes out as well, as it is inherent to the oil. Since it is very rich in minerals, hot and highly saline, it is unsuitable for human use and endemic forest and fauna systems. The contaminated water with various pollutants will hinder cultivation, destroy flora and fauna, pollute drinking water and damage other associated utilities in the southwest part of Manipur.

According to the environmentalist Dr R. K. Ranjan, the area where the two blocks lie is within the Indo-Burma mega biodiversity hotspot zone. The region is very rich in endemic species of both flora and fauna, and is the second largest riverine ecosystem of Northeast India. Its botanical and biological systems have not been studied sufficiently. Medicinal plants, dense bamboo jungles and other forest resources are very abundant sources of the sustainable development. As an alternative, an economy of such resources that also fulfils the principles of justice and sustainable development will be best argument to avoid oil and gas extraction.

It is also important to assess the extent to which the green forests that will be destroyed by the project are contributing to oxygen production and mitigating climate crisis and how much carbon emission from the extraction process and the consumption of fossil fuel from Manipur will add to global climate change and warming.

The Assessment of Ogoniland, Nigeria, by the United Nations Environment Programme (UNEP) in 2011 (Environmental Assessment of Ogoniland Report 2011) revealed the tragic history of pollution and its severe hazardous impacts on people and the environment due to oil spills and oil well fires. The Ogoniland community is exposed to petroleum hydrocarbons in the air and drinking water, sometimes at elevated concentrations. The Assessment confirmed that it will take 25–30 years to restore environmental health and reverse the damage. The UNEP recommended the creation of an 'Environmental Restoration Fund for Ogoniland', with initial capital of $1 billion with financial inputs from the oil industry operators.

The hydrocarbon extraction will ruin people's lives and the environment. The conflict along the Barak River and its downstream region due to the impending oil extraction in Manipur can be avoided or checked. The Government of India as a principal stakeholder in this conflict should re-evaluate the oil project. Reviews of the two EIAs is required since they did not fairly mention the important aspect of the impacts. It will be helpful to identify the critical threat to the water resources and its eventuality on the human population in terms of livelihood and conflict. If the Government reconsiders the voices of the peoples and civil societies it will be a favourable response towards the issue. In fact, there could be a trade-off calculation between the greater loss and smaller profit of this hydrocarbon project.

References

Bedanta K. D. and S. S. Das. 2013. An inquiry into the problem of illegal migration from Bangladesh and its impact on the security of India. Oct–Dec 2013. *International Journal of Research in Social Sciences and Humanities*. www.ijrssh.com/images/short_pdf/Dec_2013_Bedanta per cent20Kr.per cent20Dutta.pdf (Accessed on 21 September 2014).

Environmental Assessment of Ogoniland Report. 2011. 'United Nations Environment Programme'. www.unep.org/disastersandconflicts/CountryOperations/Nigeria/EnvironmentalAssessmentofOgonilandreport/tabid/54419/Default.aspx (Accessed on 17 September 2014).

Environmental Impact Assessment (EIA) for Proposed Exploratory Drilling Activity in AA-ONN-2009/2Block and AA-ONN-2009/1Block Manipur for Jubilant Oil and Gas Pvt Ltd. 29 Feb 2012. SENES Consultants India Pvt Ltd.

Gupta, A. 2009. 'Development of Barak Valley: The Question of Sustainability'. www.academia.edu/1001779/Development_of_Barak_Valley_the_Question_of_Sustainability (Accessed on 27 August 2014).

Kalita, P. 2014. UN report boosts Assam's fight against Bangla influx. *The Times of India*. 14 September 2013. http://articles.timesofindia.indiatimes.com/2013-09-14/guwahati/42061955_1_foreigners-tribunals-illegal-migrants-bangladeshi-migrants (Accessed on 21 September 2014).

'Oil and Violence in Sudan'. African Centre for Technology Studies Canadian Centre for Foreign Policy Development. 2002. www.academia.edu/4962187/Oil_and_Water_in_Sudan (Accessed on 12 September 2014).

www.banglapedia.org (Accessed on 1 August 2014).

14 The Dibang Multipurpose Project

Resistance of the Idu Mishmi

Raju Mimi

Description

Seventeen large dams are proposed to be built in the Dibang River Basin in Arunachal Pradesh, with their hydropower generating capacity ranging from 20 megawatt (MW) to 4,500 MW. The 3,000 MW Dibang Multipurpose Project is one such large dam which is proposed along the Dibang River, locally known as the Talon. It is a hydropower-cum-flood moderation scheme to be set up at Munli, the project site, which is located about 43 km from Roing, the district headquarters of the Lower Dibang Valley.

The Dibang River originates at the feet of Athu-popu, a holy place of the local Idu Mishmi people, near Keya pass (16,000 feet) on the Indo-China border in the Dibang Valley District, north of the Lower Dibang Valley District. The entire catchment area of the Dibang River is spread over two districts: Lower Dibang Valley and Dibang Valley. The region is bound on the north by China, on the east by the Lohit District and China, on the west by the Siang District of Arunachal Pradesh and on the south by Assam. The total catchment area of the river up to the dam site is 11,276 sq. km. Of this, the direct draining catchment constitutes an area of 59,811.88 ha. The height of the dam above its deepest foundation level will be 288 m, with a length of 816.3 m at the top. The Dibang dam is slated to be India's highest dam.

The Idu Mishmi people live in the upstream and downstream region of the proposed dam. Further downstream areas are inhabited by other tribal communities like the Adi (Padam), Galo and Mishing. An area of 4,009 ha will be submerged by the dam, and a back water reservoir of 43 km will be created. The project will acquire 5,827.8 ha of land, of which 5,056.50 ha is community forest land classified as Unclassed State Forest (USF).

As per the survey by the National Productivity Council (NPC), the agency entrusted to prepare an environment impact assessment study,

43 villages consisting of 859 families from the Idu Mishmi community are directly affected by the dam due to submergence and construction related activities. The total population of these households is 1,877. The impact on the downstream region has not been studied, and the number of people to be affected is yet to be ascertained.

The conflict

The hydropower potential of the State of Arunachal Pradesh is pegged at 50,000 MW – the highest in the country. The government of Arunachal Pradesh claims that if this hydropower potential is harnessed, the State could meet one-third of the power requirement of the country. In the process, the State would earn abundant hydro-dollars, similar to Arab countries which are awash with petro-dollars. To achieve this objective, more than 100 Memorandum of Understanding (MoUs) has been signed with various power developers by the State government to develop hydropower across the State.

The primary justification for the 3,000 MW Dibang Multipurpose Project (Map 14.1) is the 'revenue' the State will earn from its 12 per cent share of power in the form of water royalty. It is estimated that this project alone will earn revenue for the State to the tune of Rs. 300 crore a year. For the proponents of the project, it is 'economically viable' as it has little direct 'displacement' and negligible 'rehabilitation and resettlement' issues. This underlying justification has made the project highly controversial.

The Dibang Basin is a thinly populated region comprising of the Lower Dibang Valley and Dibang Valley districts, whose primary inhabitants are the Idu Mishmi people. According to the 2001 census, the population of the Mishmi (Idu) is around 11,023. Due to their low population, the Idu Mishmi people fear the influx of outsiders for dam building, which will lead to a demographic imbalance in the Dibang Basin (Vagholikar 2007). The proponents of the Dibang dam, the National Hydro Power Corporation (NHPC), estimate that a workforce of around 5,800 people (labour and technical staff) will be needed for project related activities. This estimate has been contested by the All Idu Mishmi Students Union (AIMSU), the student group of the Idu Mishmi community, which has a stronghold in the region. The student group claims that a single project will bring about 15,000 people into the region. For the 17 hydel schemes in the Dibang Basin, more than 100,000 people will be required to be brought in from outside. The Idu Mishmi Cultural and Literary Society (IMCLS), the apex

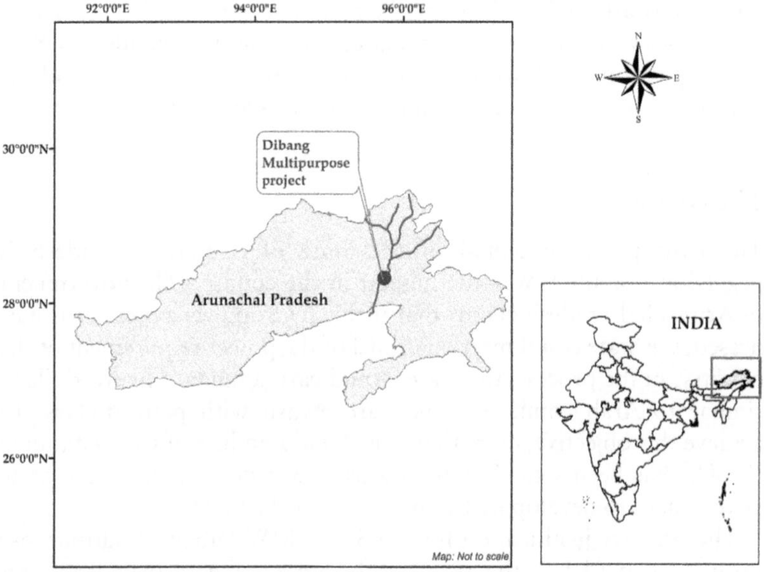

Map 14.1 Case study area
Source: SOPPECOM, Pune

body of the Idu Mishmi community, has expressed similar concerns about the viability of the project. According to this society, the claim of the proponents that the project will cause 'negligible human displacement' grossly undermines its harmful impacts on their small community. However, the State government ignored their concern as 'mere fear in the minds of the people'. This attitude of the State government led to a series of conflicts between the state and the community.

The opposing stand and the community's resistance to the Dibang Multipurpose Project

The onslaught of several hydropower projects on this river, revered by Idu Mishmis, stimulated a debate about the right to ancestral land, identity and culture. For the Idu Mishmi people living along the tributaries of the Talon River, these projects spelled 'cultural genocide'. The local people who are against hydro projects, led by the IMCLS and the AIMSU, do not believe that the construction of the dam will bring prosperity. They believe, instead, that it will deprive them of whatever they have, and that it is premature to construct a mega project in the district. They fear that once the construction of the dam begins, they will lose their political rights – rights over their land and natural resources – because the legal mechanism to prevent the inflow of migrants into the region will fail.

The IMCLS and AIMSU jointly took on the responsibility of resisting the imposition of dams on their community land. They opposed the first public hearing for the Dibang project scheduled on 23 and 25 May 2007 at Roing (Lower Dibang Valley) and Arzoo (Dibang Valley) respectively. The Government of Arunachal Pradesh then issued a notice on 4 May 2007 inviting expressions of interest/bids for the implementation of nine more mega hydroelectric projects on the Dibang River. This led to resentment among the people, and on 28 May 2007, in a large public meeting in Roing which was attended by the local legislator, the IMCLS and AIMSU took the stand that no dams should be built in the Dibang Basin. Through litigation and petitions, the organisations managed to postpone the dates for the public hearings to 30 June and 3 July 2007.

In response to the growing public resentment, three legislators from the Dibang Basin organised a pre-hearing meeting on 14 June 2007 in Roing, under the banner of the Dibang Basin Welfare Committee (DBWC). The public lambasted all the three local legislators present at the meeting for attempting to facilitate an early hearing instead of raising public awareness about the issue. The three legislators then agreed to postpone the public hearing to the last week of November 2007.

The AIMSU then carried the protest movement further to the Dibang Valley District, initiating a wide public debate on dams. In a meeting held on 3 April 2007 at Anini, the district headquarters of Dibang Valley, the local people voiced their opposition to dams and condemned the government for disregarding people's concerns. The meeting was attended by the local legislator, panchayat leaders and the Deputy Commissioner of Anini. Thereafter, the AIMSU issued a wide press statement de-recognising the DBWC constituted by local legislators.

As public opposition against dams gained momentum, the State government received an advance payment of Rs. 225 crore from the NHPC for the Dibang Project on 25 June 2007. The State government justified the payment claiming that it was meant to revive the ailing Arunachal Pradesh Cooperative Apex Bank. The AIMSU opposed the payment by declaring a 12-hour bandh on 2 July 2007 in the two districts – Lower Dibang and Dibang Valley. Despite the opposition, the State government issued a notice for the public hearing, which was scheduled on 14 and 17 November 2007 at Roing and New Anaya respectively.

Meanwhile, the Dam Affected Citizens Committee (DACC), an organisation comprising local political leaders from project affected families of the Dibang project, started negotiating with the State government. The State government postponed the November public hearing to make way for the meeting with the DACC. The meeting took place in the Chief Minister's chamber at Itanagar, the State capital of Arunachal Pradesh, on 28 November 2007. It was attended by DACC members, NHPC officials and State government officials. The State government ordered that all the project affected sites be surveyed again to accurately incorporate the losses in the EIA document of the Dibang project.

The AIMSU chose to stand by its position of 'no dams'. It opposed the resurvey, and on 21 January 2008, it held a public gathering in New Anaya, a project affected village. At this meeting, the people consented to opposing the resurvey. A signature campaign took place through which the affected people conveyed their message of 'no to dams' and hence 'no to the resurvey'.

One month later, on 29 January 2008, the first public hearing for the Dibang Multipurpose Project took place in Roing. The AIMSU carried out a poster campaign opposing the dam on the day of the hearing. The public hearing was a 12-hour marathon session from 10 am to 10.30 pm. The second public hearing was scheduled on 31 January 2008 at New Anaya. On that same day, the Prime Minister laid the foundation stone for the Dibang project at Itanagar, 500 km away from the project site. But the New Anaya public hearing could not be conducted as road communication was disrupted due to heavy landslides and snowfall on the way to the public hearing site located 130 km away from Roing.

The AIMSU condemned the laying of the foundation stone. According to the union, it was absurd that the foundation stone was laid 500 km away from the actual project site. Also, the government had acted in complete disregard of the public consultation process that was yet to be completed. On 1 February 2008, the AIMSU held a protest rally in Roing town, criticising the government for subverting its own laws. The laying of the foundation stone reflected the flaws in the environmental clearance process. It became clear to the people of Dibang that the public hearing was a mere procedural requirement rather than a legitimate process. As a result, the AIMSU resolved that no dam would be built and reiterated their opposition to the public hearing (Figure 14.1).

Figure 14.1 AIMSU protest against laying of foundation stone, 1 February 2008
Source: Jiko Linggi

The second public hearing for the Dibang Project, which was cancelled earlier, was rescheduled at New Anaya on 12 March 2008. On the day of the hearing, the AIMSU organised the local affected villagers to stage a road blockade 15 km away from the site of the public hearing along the Roing-Anini highway. The government and NHPC officials who came to conduct the public hearing were stopped midway and were forced to turn back (Figure 14.2).

On 18 July 2008, the AIMSU organised an interstate dialogue with the All Assam Students Union (AASU) at Roing, calling for the scrapping of the Dibang Project. The Downstream Anti Dam Committee (DADC) represented by the downstream Adi people, DACC and IMCLS also participated in the meeting (Figure 14.3).

Having failed for the sixth time in conducting a public hearing for the Dibang Multipurpose Project, the State government announced that the second public hearing would be held on 20 August 2008 at Anini. This time, the hearing was scheduled in the district headquarters of Dibang Valley, approximately 100 km from the actual hearing site, New Anaya. The AIMSU led an aggressive campaign to go ahead with its demand that no dam should be built. It threatened to close down

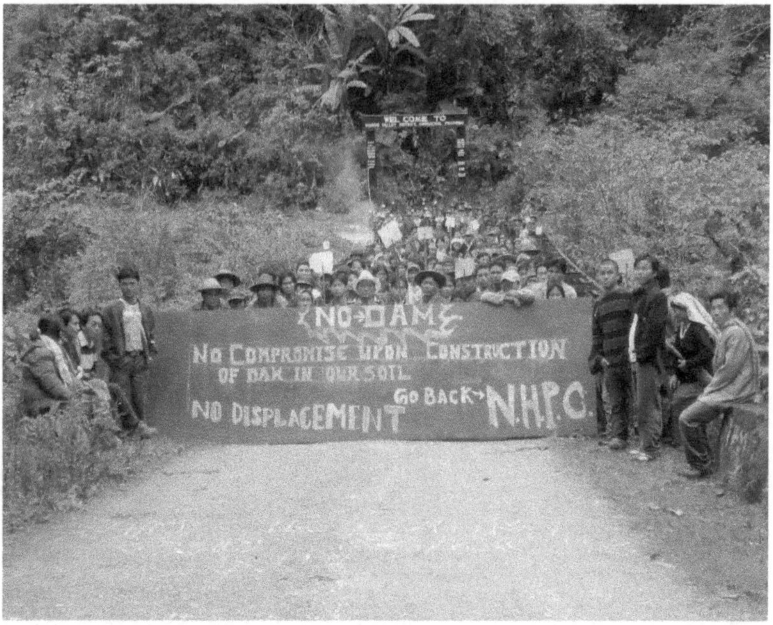

Figure 14.2 Dibang hearing boycott, 12 March 2008

Source: Jiko Linggi

Figure 14.3 Interstate dialogue for scrapping Dibang project, 18 July 2008

Source: Raju Mimi

the office of the NHPC at Roing. On 11 August 2008, a 12-hour dawn to dusk *bandh* was observed in the two districts of Lower Dibang Valley and Dibang Valley. During the strike, the AIMSU mobilised support of the DADC along with that of the Takam Mishing Poring Kebang (TMPK) and the Sadiya Mohokma Suraksha Samiti (SMSS) of Assam. A large number of protestors gathered outside the NHPC office and threatened to close it down. They were stopped by a strong police force. Responding to the protest, the Chief Minister expressed his willingness to talk with the organisations that were opposing the dam. Meanwhile, the public hearing fixed at Anini was cancelled following a legal notice issued in violation of the EIA notification.

The talks with the Chief Minister were scheduled for 28 August 2008 at Itanagar, and the public hearing was scheduled on 23 September 2008 at New Anaya. The AIMSU in a press statement called for a boycott of the meeting and the public hearing. According to the union, the meeting was just a means to pacify the protestors so that the project could proceed. It resented that the government had announced the public hearing even before the meeting with the Chief Minister had taken place. The union, however, failed to draw the support of the DADC and DACC, and had to attend the meeting on its own. On the day of the meeting, the AIMSU made its stand of no dams clear, and as a result the talks failed.

Meanwhile, new organisations of project affected people emerged. Responding to an appeal from the Dam Affected Committee and the Middle Dibang Valley Students Union (MDVSU) for talks, the government cancelled the public hearing scheduled on 23 September. However, the talks could not be held because the DACC and AIMSU opposed them on the grounds that the affected people were not in favour of holding them. Regardless, the government scheduled the public hearing on 25 November 2008. In response, the AIMSU declared a 12-hour dawn to dusk *bandh* on 19 November 2008 in the two districts. It also staged road blockades at various locations along the Roing-Anini highway. After a week of protest, the government postponed the public hearing.

After the public hearing was cancelled for the ninth time, the government rescheduled it for 27 March 2009. The hearing was again cancelled after several organisations including the AIMSU, the Dibang Valley Students Union (DVSU) and the DACC declared a road blockade along the Roing-Anini highway from 25 March till the day of the hearing.

Meanwhile, the grant of environmental clearance for the Dibang project was held up for another two years as the NHPC was required to carry out preliminary activities pertaining to the clearance in the wake of massive agitations launched by the local population against the project. Accordingly, in a meeting on 16 June 2009, the Expert Appraisal Committee (EAC) directed NHPC to take up additional studies for the EIA and conduct a public hearing in Roing and New Anaya afresh.

The impacts of the conflict

Ever since the proposal for a large number of hydropower projects in the Dibang Basin was mooted, and opposed by the affected people relentlessly, the State government started mobilising paramilitary forces in the region. After the first public hearing was declared in 2007, two companies of the Central Reserve Police Force (CRPF) have been permanently stationed at Roing, besides additional forces of the State armed police battalion and the Indo-Tibetan Border Police (ITBP).

Also, the government of Arunachal Pradesh planned to deploy 20,000 police personnel for the interstate (Trans-Arunachal) highway, and 25,000 police personnel for the various hydropower projects being built in different parts of the State.

In the Lower Dibang Valley, the presence of heavy paramilitary forces gave rise to occasional tension between the police and the people. On 25 December 2009, a group of CRPF personnel went berserk, assaulting the civilian population in a busy market area of Roing town. A skirmish between a *jawan* and a local citizen triggered the assault. In another incident on 26 April 2010, an elderly man from the Idu Mishmi tribe was beaten to death by a group of India Reserve Battalion (IRBn) personnel. This resulted in a major backlash from the local population, with angry protestors attacking local administrative buildings and police stations.

The conflict between the state and the community had a deep impact on the psyche of the local Idu Mishmi people. A sense of insecurity prevailed, along with a realisation that the paramilitary forces were being used against them. Local politicians used the paramilitary forces to reassert their social and political control in the region. As a result, the police began to intervene in people's personal lives, sometimes acting as 'moral guardians'. The police excesses continued and ultimately culminated in the police firing of 5 October 2011 where eight school children including a girl were shot and wounded during the Durga Puja celebration at Roing (Figure 14.4).

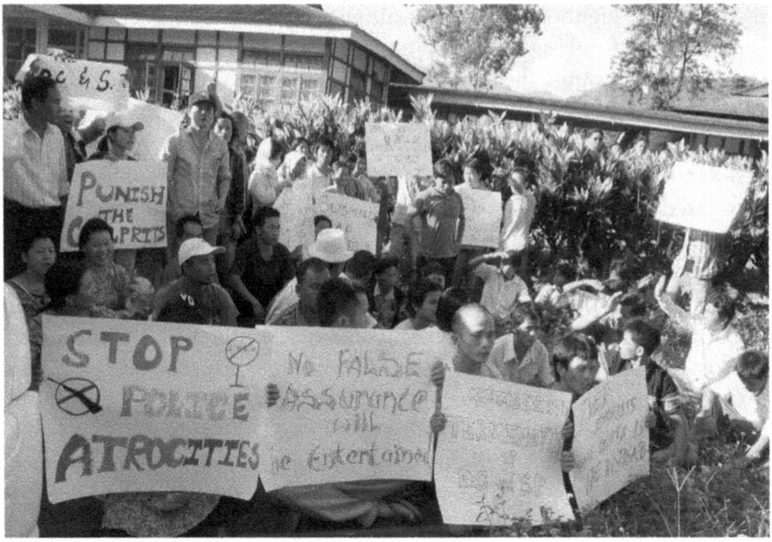

Figure 14.4 Public protest against 5 October police firing, 6 October 2011
Source: Raju Mimi

Highest points

The two-year period from July 2009 to the beginning of 2011 saw a brief lull in the anti-dam movement. With the arrival of 2011, the media and intelligence were abuzz with reports of Maoists trying to establish networks in the Northeast. According to one report of the Ministry of Home Affairs (MHA): 'The Maoists have already set up a network in Arunachal Pradesh, mainly in Dibang Valley district that borders China. Their focus as of now was to build an anti-dam opinion in the district and they are working as anti-dam activists.' In some academic circles as well, scholars stated that peaceful Arunachal Pradesh was emerging as a hot bed of Maoist activity, with people in the Dibang Valley region protesting against the Dibang Multipurpose Project at New Anaya.

With such developments, apprehensions were high that the state may deploy its military might against the local community on the pretext of Maoist activity, so that it could go ahead with the hydro project. The conflict between the state and the community was hence aggravated. The presence of Maoists in the Dibang Valley was also alleged by the local administration in Roing through their press statement of 24 September 2011. In the press statement, the administration appealed to the citizens to be vigilant, and claimed that Maoist groups may have penetrated the district from neighbouring Assam, and may try to create insecurity among the people in the area with an objective to disturb the public hearing process. The fresh public hearing for the Dibang project was to be held in New Anaya on 28 October 2011 and in Roing on 31 October 2011.

The local Idu Mishmi people were worried that the community's resistance against the Dibang dam could be labelled as a movement instigated by the Maoists. They feared it might then invite state repression. As a result, in a meeting held on 2 October 2011, the IMCLS, AIMSU, panchayat leaders, student leaders and community elders jointly accused the government of fabricating stories of Maoists in the Dibang Valley. The Lower Dibang Valley District Zilla Chairperson has on record stated that the Maoist threat has been fabricated by the district administration to mobilise additional troops so that the public hearing could be conducted by force.

The mistrust between the civilian administration and the common people reached an unprecedented level on 5 October 2011, when the police fired on eight students in Roing during the Durga Puja celebration (Overdorf 2012). The administration claimed that the students became unruly, and that the magistrate on duty had to resort to firing to control the situation. The AIMSU and IMCLS, however, have a different position on the issue. According to them, the shootout was a drill to engender a fear psychosis among the people, to pave the way for the administration to conduct the public hearing in the last week of October. The AIMSU and IMCLS claimed that the administration only needed a pretext to open fire. During this period, the conflict was at its highest point, and the entire community stood the risk of being victimised by the state machinery.

Lowest point and current status

In order to bring about a solution to the unending conflict, and to avoid any possible state atrocities on the community, the IMCLS facilitated a wide public debate for initiating a dialogue with the government. It formed a committee, the Dibang River (Talon) Basin Consultative Committee (DBCC), to negotiate with the state government. With this development, the apex body stepped back on its stand of absolute opposition to the dam. The IMCLS held the view that the deadlock that had arisen because of the conflict between the state and the community must be broken. Accordingly, to resolve the conflict and to carry the dialogue forward, it proposed a model whereby the revenue generated by the hydropower projects would be shared with the community. A meeting was held with the Chief Minister on 17 January 2012 at Itanagar to this effect, which was attended by the IMCLS, AIMSU, panchayat and public leaders from the community. During the meeting, the apex body decided to keep aside other concerns like sociocultural and environmental impacts of the dams for future discussions.

For now, the opposition to the Dibang project has been put on hold and talks are in progress. The community has proposed a mechanism whereby it can avail of a share in the revenues generated by the project. The community feels that such an arrangement, if worked out, will give them a sense of participation in the project. The State government has made a positive gesture towards the proposal. It has promised to work out a way to place it up before the assembly for legislation. This action of the IMCLS has invited criticism from some sections of the community, who feel that the organisation has sold out. However, the IMCLS considers itself incapable of disputing the developmental designs of the State any further.

Scope for dialogue

Representatives in the ruling Congress party are supportive of the IMCLS proposal for a revenue sharing model between the state and the local community. They hold the view that instead of bulldozing public opposition to the mega project, the revenue generated by the project should not only benefit the project proponent and the government, but also be shared by the people who are going to be affected, so that they can enjoy the fruits of development. One Congress parliamentarian from the State has suggested a modality similar to the Indian Mineral Development Policy, 2012, with provisions to directly include the affected people in the revenue sharing model.

Given the support of the present State government to the IMCLS proposal, it is likely that the five years of conflict over the Dibang Multipurpose Project will be resolved. However, issues like population influx and the resulting demographic imbalance leading to sociocultural and environmental impacts, which were at the root of the conflict, are yet to be addressed. It may not be a win-win situation for either party if these root concerns are not addressed.

The State of Arunachal Pradesh is declared as a protected area under the Bengal Eastern Frontier Regulation, 1873. Under the provisions of this act, outsiders are not permitted to acquire land and enter the State without a valid document issued by the government of Arunachal Pradesh. However, the current developmental initiative to build hydropower projects across the State requires unhindered movement of men and materials, which in due time will weaken the constitutional and legal protection the local people enjoyed so far. It is now left to the people and the government to collectively work upon these issues. Also, the people must decide if it is morally acceptable for them to accommodate some risks to enjoy the fruits of development.

Also, the local tribal communities are intricately linked with their surrounding natural environment. They consider the forests, rivers and land as their community resources, over which they have natural rights. However, such community ownership is not recognised legally. The State Resettlement and Rehabilitation Policy of 2008 only offers monetary compensation to the people for these resources. Such limitations can be corrected if the Gram Sabha is empowered to make decisions to manage and protect community lands.

References

Editorial. 2011. Development with brutality. *Economic & Political Weekly*, 46(42), 15 October 2011: 7–8.

Overdorf, J. 2012. Cultures in danger: 150 dams proposed in Arunachal Pradesh could devastate the state's vibrant indigenous cultures. *Global Post*. 23 April 2012. http://mobile.globalpost.com/dispatch/news/regions/asia-pacific/india/120406/dam-nation-arunachal-pradesh-indigenous-culture-part-3.

Vagholikar, N. 2007. A flood of dams. *Himal Southasian*. September 2007. http://himalmag.com/component/content/article/1277-.html.

15 Tipaimukh High Dam on the Barak River

Conflicting land and people

R. K. Ranjan Singh

The Barak River and its associated river basin ecosystem and civilisation originated from the natural lofty terrains of the Houngdu range (2,500 m above sea level) of Liyai Khullel (25°26′ 26.448″ N and 94°17′ 20.112″ E) of the Senapati District of Manipur, and extended up to the deltaic region of the Bay of Bengal. The Barak, a transboundary river located between the Indian States in the region and Bangladesh, is the second biggest drainage system in the Northeast. The river is known by different local names in the ethnic communities inhabiting the banks. It is called *Vouri* by the Pokmai, *Ahu phuak* by the Zeliangrong and *Rounglevaisuo* by the Hmar communities. The Barak regulates a diverse tangible and intangible cultural heritage of the indigenous communities inhabiting the region through which it flows. The Barak Basin serves as a cradle for varied and unique endemic flora and fauna of the region, and as a result, is a part of the Indo-Burma mega biodiversity hot-spot zone. The river has a total course length of 1,030 km, of which about 500 km drains into the geographic region of Manipur, Mizoram, Assam and Meghalaya. It has a drainage basin area of about 52,000 sq. km in India and 12,000 sq. km in Bangladesh (total 64,000 sq. km). The main stream originates from Liyai Khullel of the Senapati District of Manipur. The upper stream catchment areas are spread all over the western half of the State covering the inhabited areas of the indigenous people of Senapati, Tamenglong, Churachandpur districts and that of the Jiribam sub-division.

The flood-plain and the middle course of the Barak are located within the Cachar, Karimganj and Hilakandi districts of Lower Assam. Its important right-bank tributaries include Makru, Jiri, Labak, Mandhura, Dalu, Jatinga and Larang. The major left-bank tributaries are Irang, Tuivai, Sonai, Rukni, Katakhal, Dhaleswari, Singla and Langla. The lower course of the Barak River begins in Bangladesh near Amalshid of Sylhet District and then bifurcates. The right-hand

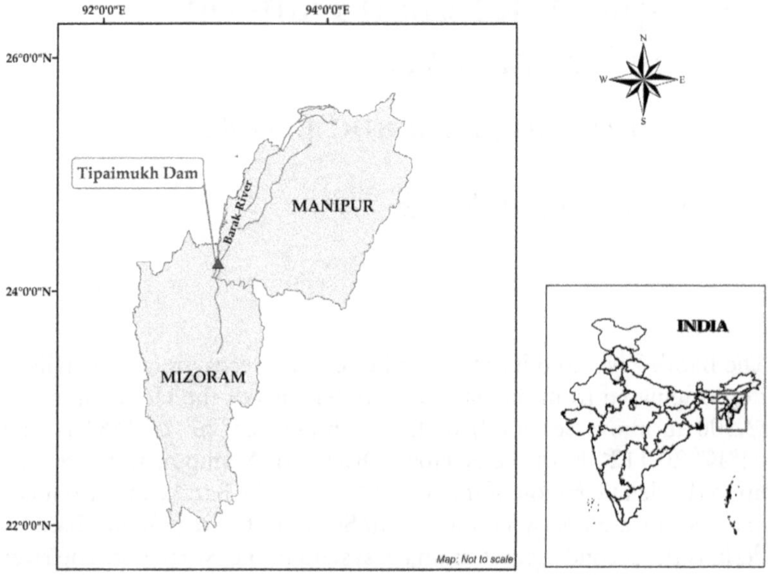

Map 15.1 Location of the Tipaimukh dam
Source: SOPPECOM, Pune

branch is the Surma River, and the left branch is the Kushiara River. The Surma and the Kushiara rivers reunite near Markuli in Habiganj District to become the Kalni River. Near Bajitpur of Kishoreganj District, this river meets the Ghorautra River and is called the Meghna thereafter. Thus, the Barak River originating in Bangladesh flows bearing the names Kushiara, Kalni and Meghna for about 530 km, before it finally meets the Bay of Bengal. The entire riverine ecosystem of the basin supports an endemic agrarian civilisation thriving on biodiversity-based agro-ecological systems that have profound local and global significance.

Emergence of Tipaimukh and its conflicts

In an attempt to control frequent flooding in the natural floodplain areas of the lower Barak plain, several proposals for harnessing the Barak River have been raised within government and political circles since India's pre-independence days. In 1954, the Assam government requested the Central Water Commission (CWC) and the planning commission of the union government to identify a suitable location where monsoon waters of the Barak could be impounded to form an artificial flooding zone (Brahmaputra Flood Control Board 1984). Accordingly, the North Eastern Council (NEC) entrusted the investigation work to the CWC. The CWC submitted its report in 1984, which proposed the construction of the Tipaimukh High Dam (THD) (Map 15.1) at a cost of Rs. 1,078 crore (WAPCOS 1989). However, the report was turned down due to the lack of a proper environmental impact assessment (EIA) of the submergible areas. Again, in 1995, at the request of the NEC, the Brahmaputra Flood Control Board prepared the detailed project report. There was no progress after this. Finally, in 1999, the Brahmaputra Flood Control Board handed over the project to the North Eastern Electric Power Corporation Limited (NEEPCO). On 18 January 2003, the project received the all-important notification under section 29 of the Electricity Act.[1]

Main features of the Tipaimukh High Dam

The project envisioned a 390 m long, 162 m high earthen-rock filled dam across the Barak, 500 m downstream after the confluence of the Tuivai tributary and the Barak on the Manipur-Mizoram border. The dam will be at an altitude of about 180 m above mean sea level, with a maximum reservoir level of 178 m. It was originally designed to contain flood waters in the lower Barak Valley, but the component of hydropower generation was later incorporated into the project. It will have an installed capacity of 6 × 250 = 1,500 MW, and a firm generation of only 412 MW. The dam will permanently submerge an area of 275.50 sq. km (NEEPCO 1998) and is feared to have negative impacts over an area of 9,126 sq. km in the State of Manipur alone. A large number of indigenous communities, mostly belonging to the Zeliangrong and Hmar peoples, will be permanently displaced and deprived of their livelihoods.

History of resistance

Though there was no comprehensive study that focused on biodiversity, environment, health, human rights, socio-economic and hydrological impacts of the proposed project and geo-tectonic problems, the communities from Manipur have resisted the THD for more than 15 years. The absence of meaningful consultation (that could lead to free, prior and informed consent) with the indigenous people contradicts the keystone of strategic priority developed by the World Commission on

Dams, that no dam should be built without the demonstrable acceptance of the affected people, and without the free, prior, informed consent of indigenous peoples as also outlined in the UN Declaration on the Rights of Indigenous Peoples. Now the struggle against THD is no longer confined to Manipur alone, but has also spread in the downstream areas including the lower Barak Valley and Bangladesh, where the immediate negative impacts of the dam are feared to be felt. The dam may also impact watersheds and ecosystems in northwestern Burma. As learnt from the experience with other projects, dams have already created or exacerbated ethnic conflicts. In the case of Tipaimukh as well, there are already divisions along ethnic lines that can have long-term implications for everyone. In an already fractured society such as the Northeast, it is imperative that the State does not allow projects that widen the ethnic divide. In addition, conflict with other States and with Bangladesh due to the dam cannot be ruled out and will need to be addressed before deciding on the project. Further, there is no clarity about its scientific and technical feasibility, EIA, rehabilitation policy and safety for downstream communities, which has created enough ground for a major conflict and raging controversy around the THD.

Conflicting geological and seismic factors

The proposed THD site and its adjoining areas are predominantly composed of the Surma group of rocks characterised by folds and faults with a regional strike of North-north-east–South-southwest (NNE-SSW) (Ibotombi 2007). The entire locality has well-developed fractures and hidden faults called blind thrusts. These thrusts could be potential earthquake foci (Ibotombi 2000). Also, the course of the Barak River opposite the Tuivai River itself is controlled by the Barak-Makru Thrust. The entire drainage basin of the Barak is littered with fault lines that control the courses of the river and its tributaries. The proposed THD dam axis is located on the Taithu fault (24°14′N and 93°1.3′ E approximately). Such faults are potentially active and may be the foci and/or epicentres for future earthquakes.

The plate kinematics of the region is very active (Ibotombi 2007). Boundary interaction (seduction zone) between the Indian and Burmese plates makes the entire region highly seismically active. Northeast India is one of the most earthquake prone areas in the world. Earthquake epicentres of magnitude 6M and above have been observed during the last 200 years (Verma and Kumar 1994). Within a 100 km radius of Tipaimukh, two earthquakes of +7M magnitude have taken place in the last 150 years. The epicentre of the last one, in the year 1937, was at an aerial distance of about 75 km from the dam site in an east–north-east direction.

Another important aspect of seismic activity is that shallow earthquakes are far more disastrous than the deeper ones even if the magnitude is relatively low (Ibotombi 2007). The majority of earthquakes that take place on the western side of Manipur are shallow (up to 50 km focal depth) due to the nature of the tectonic setting of the Indo-Burma range. Under these circumstances, the wisdom of constructing a high dam needs to be thoroughly discussed and investigated.

After the THD is constructed, the Barak Basin will be transformed, with different scenarios emerging in the upstream and downstream of the dam. In the upstream, there will be two large artificial water reservoirs; one jointly on the Barak and Irang rivers towards north, the other on the Tuivai River to the south and east. The Tuivai River curls on its alignment and has its catchment area in Mizoram, Manipur and Myanmar (Burma).

Conflicting environmental impacts

The project report (1984) states that as per the Botanical Survey of India, there is no threat to any endangered plant, and that they have not come across any rare endemic taxa or species of aquatic plants during their survey. The same report also states that as per the Zoological Survey of India, there are no endemic and endangered fauna in the area. The references relating to the flora and fauna in the proposed THD area do not seem to be based on factual and authentic field information and have been contested. The Barak Basin, along with the rest of Northeast India, is part of the sensitive Indo-Burma biodiversity hotspot, identified on account of its gene pool of endemic plants, animal species, and microbes. The absence of important information about the biodiversity of the region in the project report apparently reflects a sense of deliberate negligence and lack of seriousness in carrying out the EIA of both the upstream and the downstream areas. The Barak Basin drains a region that the international scientific community has acknowledged for its bio-researches, and as such, is of immense significance for the country and, indeed, the planet. Today the world is seriously concerned about the impact of global climate change, and the percentage of emission of methane gases by the creation of artificial water reservoirs for big dams.[2] It implies that dammed reservoirs are the largest single source of human-caused methane emission, contributing a quarter of these emissions. A total of 25,822 hectares of forest land of Manipur will be affected by the Tipaimukh dam which will lead to the felling of 7.8 million trees, and such action will seriously contribute to climate change, both locally and globally due to the destruction of the absorption capacity of greenhouse gases and also due to emission by the proposed reversion of the dam.

According to the Department of Forest, Government of Manipur, five species of hornbill are reported from the proposed THD area: the Great Indian hornbill, the Indian Pied or Lesser Pied Hornbill, the Wreathed Hornbill, the Brown-blacked hornbill and the Rufous-necked Hornbill. The prime hornbill habited area in the Tipaimukh region is located just above the sharp south-north 'U' turn in the Barak River (where the river bends sharply north from Tipaimukh in Churachandpur District to enter the Jiribam sub-division). The dam site is located exactly at this sharp bend.

The project report insists that there are no migratory birds in the area, even though the site falls on the route of several migratory species such as the *Amur* Falcon and the *Sarus* Crane. Further, it is also reported that there are evidences of the presence of the Royal Bengal Tiger in the Barak Basin.

The project report makes no mention of national parks or sanctuaries in the submergence zone, while there are, in fact, two very important wildlife sanctuaries i.e. Kailam and Bunning. Further, there are a number of wetlands unique in nature, as well as several waterfalls in the submergence zone. The Barak Basin area is one of the most important bird areas (IBAs) in the Sino-Himalayan temperate forest, the Sino-Himalayan subtropical forest and the Indo-Chinese tropical moist forest. More than 200 endemic fish species have been recorded from the Barak drainage system. Over and above this, the wetland complex and its biodiversity catering to the continuation of the civilisation in the Sylhet District of Bangladesh should not be overlooked. Given these loopholes, it is obvious that these sensitive issues have generated a series of conflicts related to different aspects of the proposed THD.

The project report has been prepared without conducting serious field-based research in the basin area. Therefore, all permissions extended to the project should be rescinded, and the project authority should be severely castigated for providing incorrect information. Subsequently, the THD authority could not hold any public hearing in a systematic way to provide sufficient democratic space to the people who are to be affected if the dam is constructed.

Apprehensions in Bangladesh

The people of Bangladesh are worried about the possible changes in the flow pattern after the construction of THD. The Bangladesh side of the Joint River Commission is now pressing for a minimum environmental flow in the common rivers, and that the remaining flow can be shared between riparian nations, to withdraw or divert for irrigation and other use. They further stated that the Surma being the main river flowing through the *haor* (wetlands) areas of Sylhet Division, it feeds the lowlands with flooding during the monsoon and drains out stored water in the winter. The ground elevation has a natural slope towards the Tanguar *Haor* in Sunamganj, the surrounding areas of which are at an elevation of about 3 m on average above mean sea level. The watershed areas of the Surma River spread to the Garo, Khashi and Jaintia Hills. A dam at Tipaimukh shall restrict some flood flow towards the lower basins of the Kushiara and the Surma. Storage of river water in the reservoir of the dam to produce hydroelectricity shall augment low flows in the downstream rivers during the winter, such that the *haors* in the Barak Valley will remain filled in water till January. The downstream area of the Barak River shall have controlled flow released from the dam, and as a result, it is feared that the boro crop coverage will be reduced (Haque 2010).

Conflict about sociocultural heritage

Our land and our water are our history, inalienable receptacles of our collective memory, permanent sites of great spiritual and religious significance, and the foundation of our civilisation and life. This land and the waters are not a gift from any government, but are people's very own by inheritance and right. No one can violate the right of the people to the land and environment. These are thoughts of the indigenous communities inhabiting the upper part of the Barak River Basin. The Hmar regards Ruonglevaisuo (Tipaimukh) to be a historically important sacred site of their community, just like the Ganga for the Hindus. However, THD authorities seem to have little respect for the emotional and spiritual relationship of the indigenous communities with the Barak. Through generations, the ancestors of the modern-day Manipuris have fulfilled their roles as custodians of a land of beauty and fertility, to preserve it for the future. The project report of the proposed dam says that there is no historical monument in the reservoir area of the project. However, the people of the area still remember and are trying to maintain the historic route popularly known as *Tongjei Maril* (Old Cachar Road) running through the submerged zone. The legendary Barak waterfalls (seven stages of waterfalls) and the *Atengba pat* (a wetland complex of six small lakes along the right bank of the Barak) will also be submerged by the reservoir. There is also another small sacred river island of the Hmar community called *Thi-le-dam*, literally meaning 'death and life', in Hmar. According to Hmar religious beliefs, the island is the place where the soul of all human beings has to go first as soon as they die. This site will be lost under the rising waters of the reservoir. From this island, the soul proceeds either to paradise or to hell or comes back to earth to be reborn. All these sites and holy places will be lost by the people of Manipur, alienating the people forever from their ancestral heritage and culture and inflicting a blow to their cultural identity. The indigenous people of Manipur have inhabited the upper Barak region since time immemorial, certainly long before the idea of dams came up. They have been living in peace with their environment and leading lives of contentment, governed by an ancient and strong bond with the earth's resources: land, water and forests.

Communities have a collective right to a pattern of development that is fundamentally of their own choice. People should have a right to reject the kind of development that they do not need or consider detrimental to their well-being. Indeed, the Chairperson of the UN Committee on the Elimination of all forms of Racial Discrimination has written to the Government of India on 2 September 2011 regarding THD, that people's choice of the kind of development they want should be respected, and in particular, that the Right to Free Prior and Informed Consent of Indigenous Peoples before construction of the proposed dam should be operationalised.

Health impacts

One of the most serious and least-studied consequences of large dams are the long-term health impacts due to drastic changes in the ecological balance, displacement and loss of livelihood, sudden alterations in the demographic character of the area and movements of large numbers of people involved in construction and other activities. There is no clear indication that these factors have been considered at all in the case of THD.

The project proponent has tried to promote the idea that the project is the only avenue for local people to *develop* in terms of obtaining basic infrastructure such as a surface transport network, schools and health centres. This kind of unequivocal projection can be arguably an undemocratic, unconstitutional and even immoral argument, amounting to coercion. The state has responsibilities towards the welfare and development of the people. It cannot forego these responsibilities and compromise on basic human rights. As a precedent in government policy and development planning, this is palpably dangerous.

Hydro-dynamics of the Tipaimukh High Dam

The Barak River upstream of the dam has a catchment of about 12,000 sq. km. With an average annual rainfall of 1,500 mm, the total water accumulation will be about 18 billion M^3. However, minus the minimum annual average evapo-transpiration of 1200 mm, the available runoff shall not be more than 3.5 billion M^3. Water has to be released from the dam hitting the turbines round the year to generate electricity. In that case, the average flow shall be about 40,000 cusec. This may be more than 1,00,000 cusec in flooding months to less than 5,000 cusec in the dry months. The dynamics of a dam involves the creation of a reservoir, that is, water flow has to be stopped at the dam. This shall delay the water flow towards the downstream until the optimum water level of the reservoir is reached. The dead storage is filled in the initial stage of storing the dam water, so the delay shall be shorter in subsequent years, but shall be on a regular basis. The water stored in the rainy season shall be released in drier months, such that it will increase the flow during that period in the downstream basin. Such dynamics will alter the riverine eco-system of the Barak River Basin, particularly in the downstream areas. This is a key reason for the likelihood of a long-term conflict between India and Bangladesh on THD.

Cost-benefit analysis

The cost-benefit analysis should take into consideration the social and environmental costs as well, which has not been taken into account for this project. The technological cost-benefit considerations alone will not solve the continuing and recurring problems that can occur during the construction, after construction, and once the dam becomes operational. The cost-benefit analysis of the THD varies from the first project report to subsequent reports. The estimated cost of the dam increased from Rs. 1,097 crore to Rs. 3,000 crore, and has now crossed Rs. 15,000 crore. Therefore, the calculation of a viable cost-benefit ratio should be mandatory. It is unfortunate that there is no national rehabilitation policy in the country.[3] Further, till today, the issues of downstream impacts are missing from all EIAs. Provision of land for land and basic infrastructure amenities for the displaced population are serious and difficult problems. Considering these fundamental issues, the proposed THD will not contribute to true development of the affected land and its people. As a result, the proposed Tipaimukh dam across the Barak River is turning out to be an unending conflict-ridden project for the people.

An alternative approach

Developmental planning needs to focus on a specific zone, region or river basin for effective and holistic planning. The catchment area of a river is one of the most effective planning zones. An Upper Barak Development Authority may be established for a holistic developmental approach in the entire catchment area to enable the region to grow in a self-empowered and participatory mode within a sustainable domain. Informed participation from the affected people representing all sections of the communities in the decision-making process leading to a collective agreement is imperative to arrive at decisions that are sustainable and socially just. Such a basin-wise participatory approach will lead to a better understanding of the ecosystem functions, values and requirements, and of how community livelihoods depend on and influence them. Such issues are inseparable from sensitive projects like the THD, and it is their inclusion which will have a strong bearing on the viability of the projects.

Introspection for an alternative development model and decision-making processes about dams and its application, as recommended by the World Commission on Dams, 2000, including an options assessment for energy production, conducive to the protection of the environment and the socio-economic wellbeing of the people of Manipur would contribute enormously in reducing the conflict over land and development rights issues.

Notes

1 NEEPCO 2000. Detailed Project Report on Tipaimukh Hydro Electric (Multipurpose) Project. Shillong.

2 Four Per cent of Global Warming Due to Dams, Says New Research: Methane is a more potent heat-trapping gas than carbon dioxide, although it does not last as long in the atmosphere. One year's large dam methane emission, as estimated by Lima, have a global warming impact over 20 years equivalent to that of 7.5 billion tonnes of carbon dioxide – higher than annual carbon dioxide emissions from fossil fuel burning in the USA.

3 The National R&R Policy, 2007, was approved by the Cabinet on 11 October 2007. It was published in the Official Gazette and came into force w.e.f. 31 October 2007. It was passed by the Lok Sabha on 25 February 2009 and tabled in the Rajya Sabha on 26 February 2009. However, the bill lapsed with the dissolution of the 14th Lok Sabha.

References

Brahmaputra Flood Control Board. 1984. *Tipaimukh High Dam Project Report*. Shillong: Brahmaputra Flood Control Board.

Haque, I. M. 2010. *River & Water Issues: Perspective of Bangladesh*. Chuadanga: Anushilan, pp. 132–138.

Ibotombi, S. 2000. *Tectonic Framework and Seismicity of Manipur*. Imphal: Manipur Chapter of INTACH (Indian National Trust for Art and Cultural Heritage).

Ibotombi, S. 2007. *Tipaimukh Dam is a Geo-Tectonic Blunder of International Dimension Pt. I & II*. Post Conference Publication of International Tipaimukh Dam Conference, 30–31 December 2005, Dhaka.

NEEPCO. 1998. *A Brief Note on Tipaimukh Hydro Electric Project, 6X250 Mw*. Manipur.

NEEPCO. 2000. *Detailed Project Report on Tipaimukh Hydro Electric (Multipurpose) Project*. Shillong.

Verma, A. K. and K. Ashwani 1994. Micro earthquake Activity in the environ of proposed Tipimukh Dam Site, Manipur. *Bulletin of Indian Society of Earthquake Technology*, Paper No. 345. 31(4): 174–186.

WAPCOS (Water and Power Consultancy Services). 1989. *Report on Rehabilitation and Environmental Aspects of Tipaimukh High Dam Project, Phase-1*. New Delhi.

16 Hydropower projects on the Teesta River

Movement against mega dams in Sikkim

Tseten Lepcha

Located on the flanks of the Eastern Himalayas, Sikkim was a hereditary monarchy till 1975, when it merged with India to become the 22nd State of the country.

The State shares its borders with Nepal in the west, Bhutan in the south-east and China in the north. Sikkim is a land of dramatic contours with rugged mountains, deep valleys, dense forests, glaciers, raging rivers and lakes and is a biodiversity hotspot. The State has the steepest rise to an altitude over the shortest distance. Its climate ranges from tropical to temperate to alpine. The variety in elevation has endowed Sikkim with rich botanical wealth. The world's highest National Park, the Kangchenjunga National Park in the Kangchenjunga Biosphere Reserve (Map 16.1), is located in this region. There are over 4000 species of plants and luxuriant forests which cover 36 per cent of the land. These dense forests are populated with a variety of animals, some of which are threatened with extinction today because of changes in the ecosystem.

Sikkim has three main ethnic groups: the Lepchas, Bhutias and Nepalese. The total population is 607,688. The Nepali community consists of diverse ethnic groups and constitutes the largest percentage of the population.

The Lepchas who call themselves the Mutanchi Rong Kup, are Sikkim's earliest inhabitants. The culture, customs and traditions of the Lepchas are inextricably linked to their deep bond with nature, but the changing times and modern developments are disturbing the delicate ecosystem with which they have lived so closely over centuries.

Apart from these ethnic groups, many people from the plains hailing from different parts of the country are settled here, in addition to a small community of Tibetan exiles. In recent times, the State has witnessed a large infusion of migrant labourers brought here to work on large hydroelectric power projects like the Rangit and Teesta

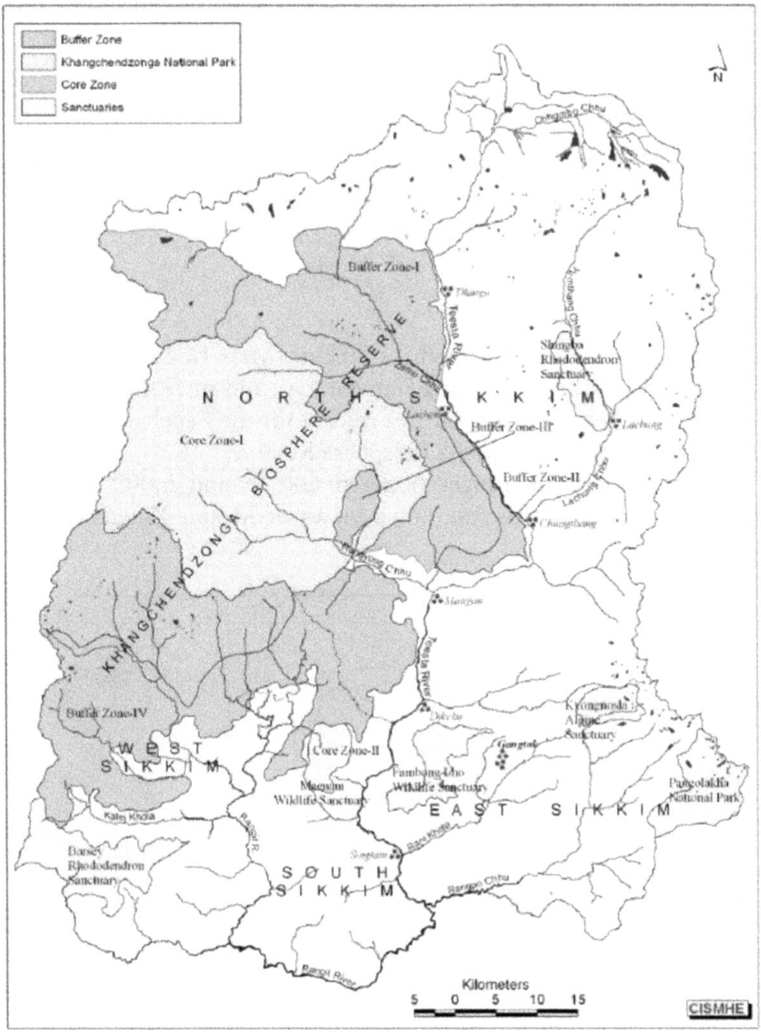

Map 16.1 Protected areas in Teesta Basin in Sikkim

Source: Centre for Inter-Disciplinary Studies of Mountain & Hill Environment (CISMHE)

Hydroelectric Project (HEP) Stage V. Also, the Border Roads Organisation (BRO) is continually bringing people into this region.

The total land mass covers 7,096 sq. km, and lies within 27° and 28° latitudes, and 88° and 89° longitudes. The crowning glory of Sikkim is

Mount Kangchenjunga, the third highest mountain in the world. For the Sikkimese, Kangchenjunga, the revered abode of their guardian deity Dzo-nga, is much more than a mere mountain.

North Sikkim

The area of the northern region of Sikkim is 4,226 sq. km, with a population of 43,350. The district headquarters is Mangan, and Chungthang is the sub-divisional headquarters.

Dzongu

Dzongu is located in the north of Sikkim (Map 16.2). The Lepchas have been living here from time immemorial, and since the 1960s, this territory has been demarcated as a reserve for the Lepcha community. It borders the Kangchenjunga Biosphere Reserve.

Dzongu's elevation ranges between 800 m and 6,000 m above sea level. The area has panoramic views of Mount Kangchenjunga

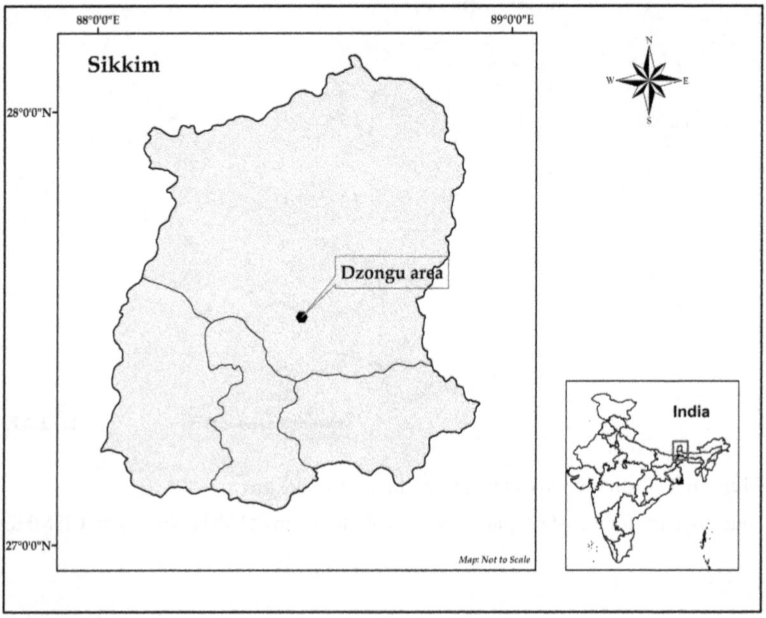

Map 16.2 Dzongu area in Sikkim

Source: SOPPECOM, Pune

(28,156 feet), and has particularly rich fauna and flora that is endemic to the area.

The Lepchas of Dzongu not only practise Buddhism, but also follow their age-old practices and beliefs, being presided over by Muns, Bhongthings and Padims. Some of them have converted to Christianity. The Lepchas of Dzongu have lived in relative isolation, primarily depending on subsistence agriculture and the cultivation of large cardamom. Unfortunately, over the past few years, the production of cardamom as well as oranges and ginger has fallen drastically. As a result of reduced incomes, younger Lepchas have migrated to urban areas, including the capital Gangtok, to pursue education and employment options, leaving behind a relatively elderly population in Dzongu.

Hydropower initiatives: the background

The Central Electricity Authority, after its 2001 preliminary ranking study of the hydroelectric potential of river basins in India, launched a hydropower initiative to produce 50,000 megawatt (MW) of electricity. Twenty-one hydroelectric power projects were identified to be developed in Sikkim.

Six projects have been envisioned on the Teesta in Sikkim, of which Stages I–IV are in north Sikkim. The Teesta HEP Stage V (510 MW) was the first to be taken up in the six stage 'cascade' plan to harness 3,635 MW of hydropower, all within 175 km of the Teesta River in Sikkim. This is located in the North and East Districts, and has been commissioned. Stage VI (500 MW) will be located further downstream in East and South Districts.

Two more projects are coming up in West Bengal, the Teesta Low Dam III and IV, which are under construction. Further, various projects are being planned on the tributaries of the Teesta in North Sikkim, with at least eight small and big projects in the Lepcha reserve of Dzongu, including the Panam HEP designed to generate 280 MW. There are three projects planned on the Lachung River (*chu*): the 99 MW Bimkyong, the 99 MW BOP HEP and the 99 MW Thangchi HEP, and numerous smaller ones. As of now, there are seven projects planned in Dzongu, including the Teesta HEP Stage III. The State government will be allocated 12 per cent of the total power generated by the projects.

It is argued that Sikkim's perennial water system should be exploited to produce cheap, plentiful power for the nation, further economic benefits through power exports, generate employment and control floods, with little direct 'displacement' of local communities. However, several unique features of the State – the geological fragility and seismic activity, the unique tribal communities and their cultural and spiritual association with river systems, their traditional natural resource-based livelihoods and the biodiversity richness of the area – pose a challenge to the conventional dam-building wisdom.

Besides, mismanagement of project implementation, general apathy towards the sentiments of the people and a total disregard for the conservation of the fragile eco-system have engendered a dislike for and fear of these projects among the people who are exploited time and again without any proper forum to redress their grievances.

In spite of this, the Government of Sikkim has issued letters of intent and signed memorandum of understanding (MoUs) with various Central government undertakings and private power producers to begin construction of these projects. It has now become imperative that some urgent measures be adopted before these destructive 'developmental projects' are permitted to be implemented.

Background of the conflict

Since the merger of Sikkim with the Indian union, the Government of India has provided many safeguards for the people of Sikkim. Article 371(F) of the Constitution is the most important instrument for the protection of the people of Sikkim. It reflects the spirit of the 8 May agreement, by way of which the old laws and traditions of Sikkim are protected. Likewise, the North District of the State, situated in a remote and inaccessible region and inhabited by aborigines, is further safeguarded by various provisions. The most prominent of these is Notification No 3069, which prohibits the settlement of non-indigenous people in the region. Further the Dzongu Area, which is inhabited primarily by the Lepchas, is a reserved area, and the region beyond Toong up to Lachen and Lachung is restricted.

However, violation of laws has diluted this measure meant for the security of the people. Further, due to the gradual migration of people owing to various projects and issues of national interest, the local people have almost been outnumbered. The increase in the population has caused serious damage to the environment and the demographic profile, culture and religion, thus affecting the lives of the poor and docile populace and also endangering their survival in their natural habitat. For instance the people of Lachen, Lachung and Chungthang region have already sacrificed 40 per cent of their good cultivatable land in the hilly terrain for the army and the BRO. In fact, some of them are almost landless. The armed forces and BRO not only brought in a large number of labourers, but also changed the names of lakes, ridges and villages extensively. For instance, the holy Guru Dongmar Lake was renamed the Guru Nanak Jheel. Further, the holy stone (*Leydo*) complex at Chungthang after which Chungthang was named, was changed into Changithang. The history of Sikkim's patron saint Guru Padmasambhava has been altered and attributed now to Guru Nanak. Meyong chu was renamed Hanuman chu, Tsolamu and Lhonak valleys are referred to as Gaigong and Kerang, Sheraythang at Chungthang as Patal Puri and another hillock as Aakash Puri, the Tashi view point at Gangtok as Patel view point.

The National Hydroelectric Power Corporation (NHPC) is responsible for the most recent cases of neglect of the local population, when it executed the Rangit River Hydroelectric Power Project and the Teesta Hydroelectric Power Project Stage V. The NHPC not only defaced the slopes of Kewzing, but also changed the name of the place to Rangit Nagar. The diversion of the river there has destroyed the biodiversity in the catchment area, introduced new diseases and caused other detrimental impacts.

In the case of Teesta Stage V, the displaced have not been rehabilitated, and jobs have been denied to local people by the NHPC. There have been blatant violations of environmental and forest laws during the implementation of this project. In fact, the State government had filed a case against the project implementers for the violations. The project has resulted in serious impacts due to tunnelling, which were not foreseen at the stage of studying the impacts of the project. Due to the indiscriminate disposal of muck into the Teesta River, the Lum Bridge which connected the Lum village in Dzongu with the region broke down.

There have also been severe health impacts at the project site. In spite of the condition laid down by the Central government that labourers should be given work permits only after a complete health check-up and treatment for diseases, this has not been followed at the State level. Therefore, it has resulted in the spread of new diseases in the project area.

Due to the instances cited above and many more similar ones caused by the carelessness of the implementing agencies and the inability of the regulatory bodies at the Central and State level to safeguard the interests of the people, fear of large projects has taken root in people's minds, which is detrimental to project implementation.

In spite of people's concerns, the Government of Sikkim has consented to build and operate 13 large hydro projects in the ecologically fragile and demographically endangered region of North Sikkim. The Government has never bothered to provide information and seek the free, prior and informed consent of the people, whose very survival is endangered by these projects. In fact the MoUs have been signed without the knowledge of the affected people, and the Sikkimese public at large. The cumulative impact of these new projects envisaged on the Teesta and its tributaries is likely to far exceed the impacts of earlier projects.

Impacts on political rights

The detrimental impact of these projects on the political rights of the people has not been addressed at all. Labourers who come to Sikkim to work on various projects continue to reside in the State as it offers further opportunities for work. This has affected the social, economic and political situation in the State, and has put great pressure on its sparse resources including land. The new entrants into the State also earn voting rights, and thereby affect the local political system.

1 The constitution provides for 12 seats reserved for the Lepcha-Bhutia community, and one for the Sangha in the 32-member State Assembly. This was done on the basis of the population ratio way back in the 1970s. Since then, due to the increase in the population of other communities, mainly due to development projects, the validity of the ratio of reservation of seats in the Sikkim Assembly is now being questioned. Any further increase in the non-indigenous voting population will result in greater marginalisation of the indigenous people of Sikkim.
2 In the reserved constituencies, the increase in the number of non-indigenous voters will defeat the very purpose of the reservation, because when non-indigenous people constitute the majority, the issues and concerns of the indigenous people are sidelined.
3 The imbalance in the percentage of voters at the grassroots level is more severe. As of now, most gram panchayats /wards have a small number of voters. Even a slight change can have serious implications on the political rights of the people at this level.

Impacts on religion: violating the scarred landscape

The local people, irrespective of their religious affiliation, are also nature worshippers. The trees, plants, mountains, streams and rocks are considered sacred by them. In fact, one of the environmental impact assessment (EIA) reports also acknowledges this. However, in some cases, the EIA report denies this fact. For example, the Teesta HEP Stage III report mentioned that 'no monument of cultural/religious/historical/archaeological importance is reported in the project as well as the study area'. This is not true because the entire natural landscape is considered sacred by the people. For example, the *Leydo* or holy stone at Chungthang along with the only paddy field here is considered to be blessed by none other than the patron saint of Sikkim, Guru Padmasambhava, in the eighth century while he was on his way to Tibet. The stone has a footprint of the Guru and a perennial source of holy water. Another example is that of the *Tamring Ney* located a little below Theng village, which the Lepchas believe is a sacred tunnel connecting Theng with *Ney Go* above Shipgyer village.

There are a number of such sites of religious significance in the study area which a proper study would have highlighted. It is a matter of great concern that the existence of these sites has been completely ignored by the project EIA report.

As has been experienced in Sikkim earlier, with the influx of a large number of people of diverse religions for mega projects, there is a sudden increase in the places of worship, which impacts the existing cultural and religious landscape and practices.

Impacts on cultural identity

With the arrival of new mega projects, there is a general tendency to proactively influence the local cultural identity through the introduction of new languages, methods of celebrating festivals, music and drama. Since the indigenous people are docile, they normally adhere to these new cultural practices. Those few who don't accept them are branded as communal or fascist. One of the studies on the Teesta HEP Stage III Project has mentioned that new communities will emerge in the project area due to the intermingling of different cultural practices.

Impacts on livelihoods due to loss of opportunities for sand collection

There is only one sand quarry in the area at present, located in the Munshithang, Chungthang region, which is an important source of sand for local construction activities and an important source of people's livelihood. After Teesta Stage III, the next upstream project to be implemented will be Teesta Stage II. The normal flow of water will be affected by this project, thereby impacting the deposition of sand in the Munshithang area and making this quarry unusable. Apart from this quarry, the closest quarry is near Yumthang, approximately 40 km away. If the Munshithang quarry can no longer be used, the cost of construction in the area will increase greatly, and the local people will be deprived of their livelihood.

Law and order problems

The environmental and social impact assessment studies do not reflect the possible law and order implications of such a mega project in a sensitive border region. Moreover, the very concept of a restricted area is defeated when such a large number of people from outside are allowed to enter and stay in the area for a long stretch of time. Forty-eight persons died during the construction of the Teesta HEP Stage V project, and more than 50 have died during the construction of the Teesta HEP Stage III project. These numbers include deaths due to earthquakes, road accidents, blasting, etc. There are numerous cases of physical assault, theft of scrap material, drunken brawls and infidelity registered in the police station in Chungthang.

Environmental issues

Teesta flowing 'underground' for long stretches

One of the most glaring environmental impacts of the hydroelectric projects planned in the Teesta Basin is that the river will be flowing underground through tunnels for long distances, leaving extensive stretches of the river course with little or no water.

This is likely to be a serious environmental hazard, impacting the riverine ecology, biodiversity and agriculture in the river valley. For example, the large cardamom plantations in the river valley, the main cash crop of the region, will get severely affected, which will destroy the economy of the region.

Impacts on agriculture

The EIA report only states the area and percentage of land in the study area under agriculture, but does not provide any details whatsoever of the kind of agriculture practised, nor of the impact of the projects on agriculture. On the contrary, it states that: 'The reduction in flow or drying of the river in the intervening stretch is not likely to have any adverse impact on the downstream users.' While direct water users of the river in the bypassed portion of the river (the tunnel area) may be few, there is no study on the impacts of the reduced flow on agricultural and horticultural lands on both flanks of the river. Crops such as cardamom require high moisture, but no information is available about the impact of the reduced flow on such crops.

Biodiversity impacts

The aquatic ecology of the river will undergo a drastic change due to this intervention in the river system. The documents of the Teesta HEP Stage III project unrealistically deny this fact by stating that 'the streams out falling between dam site and tail race discharge outfall are expected to sustain aquatic ecology'. The entire stretch between the dam and the power house will have reduced flows. The project proponents claim that there are no users of the river in this stretch. However, the river is a living entity by itself supporting several species of flora and fauna, which will be affected by the reduced flows, unless the minimum flows are calculated on the basis of the uses of the river, in consultation with the people who may be using this stretch of the river.

The EIA report admits that fish migration will be affected due to the dam construction and acknowledges that 'to prevent such impact suitable passages are provided in the form of fish ladders'. But it is not clear what measures will be implemented with respect to the impact on fish migration. Will fish ladders be provided at all? Will these be provided during the construction phase, or after commissioning? So far, there is no fish ladder in the Teesta III project, nor has any effort been made to protect the aquatic life.

The dam site of Teesta III is located very near the Kangchenjunga National Park and within the Biosphere Reserve, but neither the National Park nor the Biosphere Reserve has been mentioned in the EIA report at all. The wildlife division of the Forest Department issued a clearance certificate saying that no wildlife has been spotted in the region; however, a Serow, a schedule I animal, was found dead at the project site. The report only mentions a nameless 'Biosphere Reservoir'! The faunal base line studies are poor and do not mention the presence of important species such as the Red Panda in the study area. There is no mention of impacts on small fauna such as amphibians and reptiles in the submergence area at all. The impact on butterflies for which Sikkim is well known is not mentioned either. In the Teesta Stage V project, regulatory agencies had ordered a separate study on the impact of the project on butterflies, and after much debate a butterfly park is under construction at Thingchim.

Villages such as Theng which are to be directly impacted by the project lie in the buffer zone of the Biosphere Reserve. Quarry sites shown in the map appear very close to the boundary of the park. Even if the project components may not directly be inside the park, the heavy disturbance proposed in the buffer area needs to be studied more carefully.

Besides the environmental impacts, the disappearance of the Teesta will have a severe impact on the way of life of the people, because the folklore and ethos of local communities are intricately linked with the Teesta.

Glacial studies

River Teesta has glacial origins. The glaciers in Sikkim are located in the northern and north-western part of the Sikkim Himalaya. Apart from providing scenic beauty, they constitute the water tower in the region and control its hydrology, geo-hydrology and water ecosystem. They are the sources of all the perennial rivers and streams. Observations of the Himalaya by geoscientists have led to the detection of various rates of glacial retreat in different parts of the Himalaya. In this connection, it is observed that the Zemu glacier of North Sikkim has been retreating every year, while the behaviour of the Kangchenjhau glacier in North Sikkim is different from that of other glaciers in adjoining areas in recent times. The retreating glaciers are altering the hydrological regime in the Himalayan region and also pose environmental risks such as Glacial Lake Outburst Floods (GLOFs) and increased sedimentation. Therefore, information on the glaciers and the impact of climate change on them is critical for gauging the long-term viability of dams in the Himalayan region.

There is no information on glaciers and the environmental risks due to changing glacial behaviour on the Teesta River system in any of the EIA reports.

Floods

The rainfall in the entire catchment is drained through the Teesta River. The area is also susceptible to cloud bursts. The monsoon in the region is vigorous, adding up to a minimum of 1700–1800 cumec. It can be as high as 15,000–20,000 cumec during floods. During the 1968 floods which damaged the 80 feet high Anderson bridge, a record of 15,000+ cumec was recorded, says Dr Jeta Sankritayana of North Bengal University. Some scientists suspect that GLOFs may have been the cause of these floods. The presence of a hydrological installation in such a river can pose a serious danger to downstream areas, and the current risk assessment procedures are absent or inadequate. It is not enough to proclaim later that a 'natural disaster' hit the project. Comprehensive risk assessment needs to be carried out and made public during the planning phase.

Impacts of floods need to be studied not only after the project is commissioned, but also during the construction period. The Teesta V project experienced a huge flood a couple of years ago. The indiscriminate dumping of muck in the river course by project authorities downstream of the dam coupled with the flood resulted in severe downstream impacts, including the washing away of the Lum Bridge connecting Dzongu with Dikchu.

Loss of land

The acquisition of land for the construction of any mega project, whether belonging to private individuals or governments, invariably deprives people of their livelihood, particularly in the case of Primitive Tribes like the Lepchas. The people subsist on agricultural land, livestock rearing, fuel wood, construction of houses, etc. The fact that one-third of the available fertile and usable land has been taken away has had a tremendous impact on them. Moreover, due to a lack of knowledge about how to use the compensation money, most landowners ultimately become landless as well as penniless. To compound the problem, like most tribes elsewhere, a large portion of the money is spent on alcohol, resulting in serious health issues which the families have to face. When people are deprived of land, they lose the very basis of their livelihood.

Impacts of roads

Mega projects require long stretches of new roads. In spite of the fact that large portions of land are used and the environment is heavily impacted during construction, the impact of roads has not been given due importance. The specific impact of road construction at each place must be mentioned clearly. This is important because people have already been at the receiving end of severe impacts due to road construction activities in this geologically fragile area. Recently, Theng village has been a witness to the impacts of such activities during the construction of the Toong–Sankalang road by the BRO. The village faced serious impacts during blasting operations for road construction, and during a downpour there was a massive landslide causing great damage to property and putting several human lives at risk. The disruption of traffic because of damages to existing roads due to fresh construction hampers tourism activities, defence movement and transportation of commodities, in addition to other activities in this sensitive border area.

Location of muck disposal sites

Muck disposal has been one of the most contentious issues with the NHPC and other developers. Time and again they have been fined and action has been initiated against them.

Impacts of power transmission lines

There is absolutely no mention of the impacts of transmission lines which will be used to evacuate the power from the projects. It is important to mention the impacts due to this component, including the land proposed to be acquired for it. We understand that Teesta Urja has already signed a joint venture agreement with the Power Grid Corporation India Ltd. (PGCIL) for building transmission lines to evacuate power from the Teesta Stage III project.

Impacts of construction at 'project sites'

Project construction is the most destructive phase. The impacts include damages to the areas above and around the tunnels and at points below these tunnels, including drying up of water sources, impacts of blasting on houses and agricultural fields, use of huge machinery, use of diesel generator with output capacity up to 2 MW, noise pollution due to continuous blasting and machinery, dust pollution and damage to roads due to heavy vehicles.

Seismic impacts

Sikkim lies in a high seismic zone. Recent scientific evidence points to the fact that neo-tectonism in the Eastern Himalayas has pronounced impacts on flooding, sediment transport and the depositional characteristics of rivers and their tributaries. The basic parameters based on which hydroelectric projects are planned in the Eastern Himalayas can change due to neo-tectonism. The hydro projects were permitted without any proper studies and relevant safeguards. The most glaring evidence is seen during the 18 September 2011 earthquake which devastated Chungthang, the dam site of Teesta III, and the adjoining areas where project activities were in progress. Even if the earthquake was not triggered by the hydro project, the areas devastated by the blasting were damaged the most.

Most experts deal with the impacts of seismicity inadequately. The reports normally speak of seismicity with respect to the dam structure, but do not mention other environmental risks associated with hydroelectric projects. The dam of Teesta HEP III was not affected by the earthquake of 18 September 2011, because it was in the initial stage of construction.

A recent study by Indian Institute of Technology (IIT) Kharagpur has clearly identified Sikkim in particular as a very active seismic zone with the possibility of an earthquake of magnitude 9 on the Richter scale. While the focus is on specific projects, it is important to look at the cumulative impacts of all the projects planned in the Teesta Basin.

Latest position of hydro projects in Sikkim as of 30 May 2012

Total projects planned: 28

Total projects on the Teesta: 6 in Sikkim, 2 low dams in West Bengal

Completed: 2 (Teesta Stage V – 540 MW, and Rangit – 60 MW)

Projects shelved due to public protest: 4 in Dzongu (Ringpi, Rukel, Lingza and Rangyong), 3 in west Sikkim (Rathong chu, Lethang HEP and Thing Thing)

Projects on hold: Panam HEP – 280 MW, Teesta HEP Stage IV in Dzongu, Teesta II in Lachen, and Bimkyong, Thangchi, Bop on Lachung River.

Total Projects in the Lepcha Reserve of Dzongu: 8. The Teesta Stage V project has been commissioned but is not functioning properly due to siltation, other environmental reasons and mismanagement. Sixty per cent of the Teesta III work has been completed, but it is now continuing at a very slow pace due to the impact of the earthquake, and financial and legal problems.

The Affected Citizens of Teesta movement

Brief history

The civil society movement to fight the devastation caused by mega hydropower projects began more than two decades ago. The need for people's intervention arose after the implementation of the Teesta HEP Stage V project began.

In 1998–99, when the Memorandum of Understanding (MoU) was to be signed between the state and the NHPC for the Teesta HEP Stage III project, a citizens group called the Joint Action Committee (JAC) was very active. Now called the Affected Citizens of Teesta, this group, formed in April 1998 and comprising of citizens of East and North Sikkim, was mainly concerned about the Teesta projects, in particular, about the demographic changes that the project(s) would lead to due to the infusion of labour from outside Sikkim, whether the project(s) would generate long-term employment opportunities, and environmental impacts.

The JAC worked out a detailed list of points to be included in the MoU; however, most of them were left out. Thus, an opportunity to plan a large project with care and execute it by ensuring that the benefits of the project be maximised in favour of local people was foregone.

The JAC felt that the State should have sought more than the mandatory 12 per cent of generated power that is allocated to it free of cost. The labour permit system could have been made more systematic. A full department should have been created to oversee the implementation of all aspects of this project. At present, there are several aspects of the project that need to be monitored. A few monitoring committees have been set up but it is unclear as to what aspects each of the committees is supposed to monitor. They all function independent of one another. There is ambiguity regarding who the members of some of the committees are, and how often they need to meet to fulfil their task. Most importantly, citizens have very little information about these committees and their mandate.

The Teesta HEP Stage V project was implemented without taking into consideration any suggestions of the people. In view of this and the impending threat to people's survival, the JAC reorganised with many new younger members and renamed itself as 'Affected Citizens of Teesta (ACT)'.

The movement initially started by sending petitions to the government not to implement hydro projects in North Sikkim. Later, legal notices were issued to caution the government against possible litigation. These forms of request were not heard. A rally (march) was organised on 12 December 2008 to highlight the issues raised by the Lepchas.

Despite a formal assurance, the government took no action for many months. Hence, the ACT and supporting organisations like the Sangha of Dzongu and the Concerned Lepchas of Sikkim started a peaceful Satyagraha with a hunger strike as the main mode of protest.

The first phase of the indefinite hunger strike lasted for 65 days, and the second phase continued for 95 days. The hunger strikes, including both relay and indefinite fasts, add up to 915 days.

The second phase of the hunger strike was called off after four hydro projects were scrapped (Ringpi, Rukel, Lingza and Rangyong).

The final phase of the hunger strike was withdrawn after the Government of Sikkim invited the ACT for a dialogue to resolve the issue.

Current status

Panam HEP: This is proposed to be a 280 MW hydroelectric power project to be constructed by a Hyderabad based company named Himgiri Energy Private Limited, which is a unit of the special purpose vehicle Nagarjuna Limited. This is their first foray into hydropower generation. Land acquisition has been completed which the villagers accuse of being completely faulty. Environmental clearance has been granted. Forest clearance is still awaited and thereby the project is yet to start.

Teesta HEP Stage IV: This is a proposed 495 MW hydroelectric project which is scheduled to be constructed in the Dzongu reserved area. It is yet to receive environmental and forest clearances and so land acquisition process has not yet started, it is interesting to note that the detailed project report of Teesta HEP Stage IV was rejected by the Central Electricity Authority (CEA) due to lack of proper data on geological studies related to the project.

Opposing status

There are large number of issues, often overlapping ones, ranging from ecological to human security, which people perceive as threats. Location of the projects is a big issue both from the geological and socio-cultural perspectives for the communities inhabiting these areas. As these projects are scheduled to be located at Dzongu, a reserved area which is considered as sacred by the communities in the State, there is large-scale opposition. Moreover, this area falls under highly seismic zone. Several monasteries, springs (e.g. the Tse Tsa Chu) as well as the Kanchenjunga National Park will also be severely affected, if these projects are constructed. There will also be widespread damage to the ecology as the river will be made to pass through underground tunnel. It will not only destroy the flora and fauna of this area but will also affect cash crops cultivation in the river plains. Moreover, influx of outsiders as labourers will also make the local communities vulnerable to various risks such as health hazards, displacement, cultural shocks, etc. Above all, there is a huge scepticism about the developers of the project as many of these companies have neither the experience, technical expertise nor the wherewithal to handle projects in seismically and ecologically sensitive zones such as Sikkim.

The highs and lows in the conflict

From the activists point of view there were several highs and lows associated with the environmental movements against dam construction in Sikkim. There are several incidents and time frames when the movements attained great heights in terms of support and mobilisation, e.g. when Dawa Lepcha and Tenzing Lepcha during a *satyagraha* went on an indefinite hunger strike for more than 65 days, similarly, the mass rally on 2 October 2008 could derive lot of attention and increase the support base for the movement. The Lepchas of Darjeeling and Kalimpong conducted simultaneous fast which received widespread media attention. Even the visit by Medha Patkar and other dignitaries in support of the movement helped its cause. The two-day *dharna* held in New Delhi on 5 and 6 December 2007, along with numerous meetings with the planning commission and different ministries have been a step forward in making the demands of the movement heard at the power centre. Moreover, the rejection of the petition by the court in Sikkim, which was filed by Tashi Tsering, on behalf of the project developers, has also been a great moment for the movement. However, when the four HEPs were scrapped that were supposed to be located in Dzongu has been a moment of victory for the movement.

Similarly, there were lows associated with this long movement. When after 65 days the fast by Dawa and Tenzing were discontinued due to their fragile health, it acted as a dampener for the movement. The passing away of Smitu Kothari of Intercultural Resources and Chukie Topden who were two of the great supporters of the movement has been a loss for the struggle. The solidarity expressed by the Lepchas of Darjeeling and Kalimpong towards the environmental cause of the Lepchas in Sikkim has been painted with a different colour by the State establishment. They were victimised by the State government in the name of maintaining social harmony in Sikkim, which was used as a ploy to malign the movement. When the negotiations with the Government broke down on 25 August 2007 along with the arrest of fifty supporters in Panam accusing them of ransacking the project site, it has been a moments of despair for the movement. The case that was filed against ACT has also been a moment of desperation for the movement.

Scope of dialogue

The Members of the ACT always kept the option of engaging in a dialogue open. Moreover, the ACT movement was democratic and peaceful. The Government of Sikkim and the ACT had numerous rounds of talks during the hunger strike, until it was withdrawn on 28 September 2009. It was only after a formal invitation from the State government for talks with the ACT that the historic 915-day hunger strike was withdrawn.

However, no formal talks were held for almost seven months after the end of the hunger strike. Since the State government did not comply with its commitment, the ACT had no option but to seek legal redressal, and hence filed a petition in the high court. However, once the petition was filed, the government invited the ACT for a dialogue in April 2010. At the same time, the government, on the pretext of the matter being sub-judice, did not want to discuss various important issues, thus making the very exercise meaningless.

Since then, there has been no formal interaction with the government. At the same time, no work has been started on two contentious projects – the Panam HEP and the Teesta HEP Stage IV.

Annexures

Submission of the issue to UNESCO through the World Mountain People Association on 4 October 2010 in Paris.

The General Assembly of the World Mountain People Association wishes all United Nations Organizations to impress upon both the Government of India and the respective State Governments to fulfil the wishes of the Mountain People to live with security and dignity in their own homeland.

ENCOURAGED by the adoption of the Declaration of the Rights of the Indigenous Peoples by General Assembly of the United Nations in 2007 as the framework within which states and indigenous peoples can reconcile with their painful histories and can move forward towards the path of human rights, justice and sustainable development for all.

STRENGTHENED by the concerns expressed by the UN Committee on the Elimination of Racial Discrimination (vide CERD/C/IND/CO/19 dated 5 May 2007) at the construction of large dams in Northeast and also specific recommendation to the Government of India (GOI) to seek free prior informed consent of the communities.

COGNIZANT of the fact that 'Development' in the region is unplanned and implemented with utter disregard of our people's inalienable rights; and imposed on the basics of geopolitical and economic interests, driven by the interest of the dominant communities and corporate entities and articulated in the rhetoric of Nation interest and Welfare.

FURTHER COGNIZANT of the fact that GOI has militarized the region for more than half a century by imposing laws like the **Armed Forces Special Powers Act, 1958, based on 'National Interest'**, the eco-liberal thrust of development aggression has exponentially aggravated the process of militarization thereby proving fatal to the survival of the peoples especially in Manipur AND THAT WILL affect the entire NE region.

APPALLED by the exclusivity and lack of transparency of the state governments in entering into numerous agreements and deals with the public and private entities leading to a systematic alienation and invalidation of people's rights in the region.

WITNESS to the hollow promises of dams such as Teesta HEP Stage V in Sikkim, Loktak Project (Manipur), Doyang Project (Nagaland), Dumbur Dam Project (Tripura), Pagladia Project (Assam), Ranganadi/Partia Project (Arunachal), etc. In bringing energy security and emancipation to the lives of our people as after they are commissioned have only brought untold sufferings of widespread displacement and impoverishment, submergence of agricultural land, devastation of environment, loss and of exotic flora and fauna species, loss of identity, demographic and cultural distortion, the transfer of population from mainland India to work in these dams will substantially alter the demographic composition of the region, marginalizing the indigenous population in their own territories, will create further conflict between the indigenous population and the immigrants; thus creating an environment of unjust and painful violation of human rights.

CONCERNED that All the Laws established are violated to cater to the need of the corporate world.

The World Mountain People Association wishes that the Government of India and respective State Government should initiate immediate steps to fulfil the following demands of the indigenous people of the region.

1 SIKKIM: STOP THE CONSTRUCTION OF TEESTA HYDRO ELECTRIC POWER PROJECT STAGE IV (510 MW) AND THE PANAM HEP (300 MW) in DZONGU.

This is the only option to save the demographically endangered 'LEPCHA Tribe' with most unique and distinct culture, Tradition and language from extinction and to protect 'Dzongu' the environmentally fragile reserve which is the cradle of Lepcha civilization.

2 Meghalaya: Scrap Uranium mining project in the state in the name of development thus violating the land, forest, health and human rights of the people.
3 Arunachal Pradesh, Manipur, Assam: stop all destructive Mega Hydro electric projects.
4 Tripura: stop all forms of discrimination and land grabbing, forest and give justice to the Borok People.

Submitted with the hope that the voices of the Mountain people will be heard by the Highest institution of world and necessary action taken to ensure the rights of the people are protected to ensure a life of Security, Prosperity and Dignity.

OLORON-SAINTE-MARIE PYRENEES, FRANCE
1st OCTOBER 2010

17 An uneven flow?

Navigating downstream concerns over China's water policy

Nimmi Kurian

Understanding China's resource choices

China's growing water thirst and energy hunger is resulting in a concerted national campaign to develop and augment the country's huge hydropower capacity. China is witnessing a virtual dam-building boom in the wake of wide-ranging reforms that were introduced in the energy sector. Much of this expansion is based on augmenting capacity in its Western region, which is being projected as the energy powerhouse of the country.[1] Within the western region, Tibet is emerging as a focal point of hydropower expansion with a series of dams being planned on major international rivers such as the Salween, Mekong and the Yarlung-Tsangpo. As the headwaters of many of Asia's mighty rivers, the Tibetan 'water bank' in every sense becomes Asia's water bank and the environmental sustainability of Tibet means the environmental sustainability of much of Asia (Kurian 2004). Many of these rivers flow into some of the most populous regions of South and South East Asia.

China has begun construction of dams on the Yarlung Tsangpo including a 510 MW project in Zangmu, 640 MW project in Dagu, 320 MW in Jiacha and 510 MW Jiexu on the middle reaches of the river. Zangmu, Tibet's biggest hydroelectric project became operational in 2014. The Jiexu, Jiacha and Zangmuare are within 25 km of each other and at a distance of 550 km from the Indian border. The Ninth Report of the Inter-Ministerial Expert Group on the Brahmaputra (IMEG) called for a close monitoring of the 39 run-of-the-river projects on the Yarlung Tsangpo and its tributaries (Kumar 2013). Despite being projected as run-of-the-river projects, the fact that these will require storage of large volumes of water for power generation has raised serious concerns downstream.

Creating ripple effects

The health of Tibet's waters and the manner in which these are used upstream are bound to have ripple effects across Asia's international rivers such as the Mekong, Salween, Indus, Irrawaddy, Sutlej, Salween and the Brahmaputra. The environmental costs that get generated through the use of natural resources in a shared basin between riparian nations thus raise the potential for conflict since costs and benefits tend to flow unevenly downstream. The core issue in a shared basin becomes the criticality of perception, right or wrong. These become the vectors through which the actions of the upper riparian get refracted and processed. While technical issues of measurements, flow patterns and runoffs have their importance, it is more often the more intangible perceptual aspects that create and entrench positions that can produce or retard cooperation at the transboundary level. For instance, when runoff levels fall substantially during the lean season, levels of vulnerability among downstream communities are likely to be heightened significantly. Changes in the hydrology of the glacier-fed river also raise fears of flash floods and dam safety. There are growing concerns in the region that construction of dams by China could adversely impact flows downstream. These fears stem from the fact that India's official narrative has largely tended to downplay many of these concerns. For instance, Prime Minister Manmohan Singh summarily chose to dismiss these fears when he informed the Rajya Sabha that India 'trusts China' and relayed China's assurance that 'nothing will be done that will affect India's interest' (*The Hindu* 2011). Similarly, India's Minister of State for External Affairs E. Ahamed noted in 2013, 'India has urged China to ensure that the interests of downstream States are not harmed by any activities in upstream areas' (Ministry of External Affairs 2013). Not surprisingly, there is growing consternation in the Northeast on what is widely perceived as the Centre's lack of resolve in raising the issue with China. For instance, Assam's Chief Minister Tarun Gogoi tersely noted recently, 'Construction of big dams will adversely affect downstream areas of Assam and the north-east region and I fail to comprehend why the Centre is not taking the matter seriously despite knowing this fact' (*PTI* 2015).

As part of its response strategy, India is seeking to establish a case for user rights on the Brahmaputra by tapping the hydropower potential of Arunachal Pradesh. The moves to expedite plans to develop mega storage hydroelectric projects on the sub-basins of the Siang, Lohit and Subansiri rivers can be read as India's moves to institutionalise such a norm. But this raises a more fundamental question: what is the likelihood of India and China subscribing to the norm of prior use as a possible basis for equitable water sharing, given that both have chosen not to ratify the only existing international treaty governing shared freshwater resources. One also needs to look beyond formal and legal instruments to 'soft law' and an entire range of innovative, informal processes that allow for flexibility and consequently greater measure of success. While successful cross-border resource governance arrangements in other border regions underline this, their prospects for the South Asian region remain largely undertheorised and understudied.

If downstream communities hold the upper riparian responsible for many of their livelihood setbacks, this is bound to be a sentiment that China's public diplomacy will need to engage with seriously. This will by no means be an easy challenge for China since regional public goods cannot be expected to flow downstream in a single axis of progression from conflict to cooperation. In the absence of robust institutional backing by countries within the region for the norm of sustainable management of transboundary water resources, perceptions will tend to have a greater potential for generating conflict and mistrust. It is interesting to note the three reasons China has put forward for not pursuing the diversion of the Brahmaputra. In this regard, China's vice minister of Water Resources Jiao Yong cited technical difficulties, impact on environment and adverse impact on State-to-State relations as reasons against diversion. In addition, the prime contractor for the Zangmu hydropower project, China Huaneng Group also categorically claimed that the project would not reduce water flows downstream (Wang 2010).

China's domestic debate

The mega water projects have triggered a domestic debate within China on the viability of grandiose measures to tame the environment. A memo presented by a group of concerned scholars and experts from Sichuan questioned the wisdom of massive engineering solutions at a time of worsening climate change and glacier retreat. Ruing the lack of public debate on projects of such national significance, Professor Lin Ling, former Director of the Sichuan Academy of Social Sciences lamented that 'the planners in charge of the western route simply declared that it had been given the green light, claiming that the announcement by the Central government of the formal launch of the eastern and middle routes meant the western route could go ahead too' (ThreeGorgesProbe.org 2006). Voicing the concerns within

China's scientific and environmental community, Guo Qiaoyu, a water expert with the Nature Conservancy's Beijing office said, 'Really, you can't just change an entire environment suddenly without causing massive damage' (Simons 2006). Echoing these sentiments, Professor Liu Shiqing, Secretary General of the West Development Research Centre at the Sichuan Academy of Social Sciences, Chengdu, noted, 'We wonder whether the proposed scheme could do little or nothing to save the Yellow River, and end up destroying the Yangtze instead' (ThreeGorgesProbe.org 2006). There has also been a questioning of the likely adverse implications of a plan to divert the Yarlung Tsangpo. China's former Water Resources Minister Wang Shucheng has for instance questioned the need for the diversion project on the grounds that it was 'unnecessary, unfeasible and unscientific. There is no need for such dramatic and unscientific projects' (*Times of India* 14 October 2011). It remains to be seen if China's political leadership is inclined to show receptiveness to these concerns given increasing evidence of the negative fallouts of other mega engineering projects such as the Three Gorges.[2]

Raising missing issues

What is often overlooked in the fixation with the issue of quantity is that there are other equally critical issues of concern that need to be placed on the agenda. For instance, water quality could emerge as a key issue. Even within China, this is becoming a major concern with northern areas including Beijing anxious about the quality of southern waters that will arrive in the north. Added to these are concerns that the fragile ecosystem of the Tibet-Qinghai plateau is showing other signs of stress as it struggles to cope with the furious pace of economic activity that forms part of China's Western Development Strategy.[3] Many mega projects are transforming the face of Tibet. The 'pillar' industries of mining and timber processing have fed the rapid industrialisation of Tibet, bringing in its wake the assorted problems of deforestation, soil erosion, landslides, floods, acid rain and pollution especially of the water systems. These are creating ecological imbalances in the form of rising temperatures, retreat of glaciers and droughts caused by indifferent rainfall. Much of these will find their way to parts of the extended region.

A particular area of concern for downstream countries could be the environmental degradation facing the Tibetan 'Three Rivers area' comprising the Yarlung Tsangpo, Lhasa River and Nyangchu Basins in central Tibet, located in one of the most mineral resource-rich areas of China. One of the most intensely exploited areas in this region is the Gyama Valley with large polymetallic deposits of copper, molybdenum, gold, silver, lead and zinc. The Gyama Valley is situated south of the Lhasa River, one of the five great tributaries of the Yarlung Tsangpo that merges with it 130 km downstream. Studies by Chinese scientists are pointing to the possibility of a high content of heavy metals in the stream sediments and tailings that could pose a potential threat to downstream water users (Huang et al. 2010). Global warming could further accelerate the movement of these heavy metals besides projected spatial and temporal variations in water availability. The annual runoff in mega deltas such as the Brahmaputra and Indus is projected to decline by 14 per cent and 27 per cent respectively by 2050 (UNDP 2006). This will have significant implications for food security and social stability given the impact on climate-sensitive sectors such as agriculture.

Cumulative impact

These also raise the larger question about the cumulative impact of massive dam-building projects across the entire Himalayan region and the consequences of such intensive interventions in a region that is ecologically fragile and densely populated. The dangers of water accumulation behind dams could also induce devastating artificial earthquakes (Valdiya 1992). In the geo-dynamically active Himalayas, devastating earthquakes are an ever-present danger with a recorded history going back to the thirteenth century. A sobering reminder is that although the epicentre of the 8.6 magnitude Assam earthquake was in Rima, China, it was the Brahmaputra Valley that bore the most extensive damage. The earthquake blocked several tributaries of the Brahmaputra and created a 'wall of water' measuring 30 feet high destroying several villages in its wake. As Angus Macleod Gunn notes, the 1950 Assam earthquake despite its devastating impact, 'was not an Indian earthquake' (Gunn 2008: 417). There is also disturbing new evidence emerging about dam-induced earthquakes. Recent research by Chinese scientists has revealed that the Zipingpu Dam may have triggered the Sichuan earthquake in 2008 that resulted in 80,000 casualties (Kerr and Stone 2009).

Addressing gaps in knowledge base

Establishing a baseline database for different biodiversity and ecosystems components will be an essential prerequisite to informed public policy on sustainable development (Kurian 2014). Improving the scientific knowledge base could feed substantially into national policies for enhanced management of ecosystems. Several knowledge gaps exist with regard to glacial movements and fragmentation, wildlife species movements and changes in land use. All of these factors are crucial in determining patterns of climate change, hydrological response patterns, causes for ecological imbalances and in informing better policy decisions. Gaps in knowledge regarding transition of land use also need to be plugged as they have far reaching socio-economic as well as ecological impacts.[4] Studies in the Yunnan region have shown that in order to better monitor the changing land use, national level monitoring has been abandoned in favour of a more sub-national scale of analyses. Studies have also shown that to build these databases at sub-regional levels, cooperation of all stakeholders is required.

These only go on to underline the importance of developing a comprehensive knowledge base on water resources within the region. Disputes over data, be it of access or of accuracy can be both endemic and damaging to riparian trust building. As past crises have shown, accurate and timely information is vital for successful disaster management. This calls for evolving a system of regular exchange of data and coordination between respective national agencies in India and China as well as within the region. This fact was brought home in a tragic manner during the flash floods caused from a landslide in Tibet in 2000 that ravaged the Northeast and Himachal Pradesh. The lack of an information-sharing agreement between India and China had then resulted in an unfortunate loss of life, dislocation and extensive damage to property. Both the countries have since signed an MOU on the sharing of hydrological data on the Brahmaputra's flows, which will be vital for timely forecasting and management of floods in the Northeast. MOUs signed in 2002 and 2008 provide for information on the water level, discharge and rainfall from three stations on the Brahmaputra, namely, Nugesha, Yangcun and Nuxia from 1 June to 15 October every year.[5] As per an MOU signed by India and China in October 2013, China agreed to extend the period for provision of hydrological data to begin from 15 May, in keeping with the early onset of monsoon in Northeast India. The two countries have also set up the Joint Expert Level Mechanism, a decision taken during Hu Jintao's visit to India in 2006.[6] The Implementation Plan on provision of hydrological information signed by India and China in 2014 also includes the provision for site visits. According to Article 12 of the Implementation Plan, 'In order to ensure normal provision of hydrological information, if necessary, after mutual consultation through diplomatic channel, the parties may dispatch hydrological experts to each other's country to conduct study tour according to the principle of reciprocity' (Ministry of External Affairs 2014).

Bringing the border region back in

Many of these questions will require developing effective inter-agency coordination mechanisms among federal, State and local agencies. Since border regions are at the receiving end of a whole host of environmental challenges, capacity-building efforts should engage with and build on the experiential knowledge and coping systems specific to a region. But there are currently no platforms to institutionalise regular interactions between the Centre and the Northeastern States. For instance, the Interstate Council (ISC), a forum designed to bring all Chief Ministers to work on operationalising coordination mechanisms between the Centre and the States has only held two meetings so far, the last one being held in December 2006 (Pande 2014). These have resulted in deep-seated grievances among the Northeastern States. Recently, the Arunachal Pradesh Chief Minister Nabam Tuki vented his frustration at not being granted an appointment to meet the Prime Minister even after a month.

Centralised agencies such as the Brahmaputra Board under the Ministry of Water Resources have also shown the limits of a top-down model that does not seek to involve active cooperation by the States. These have chosen not to integrate extant traditional governance institutions and have thus virtually functioned as parallel bodies. There are also as yet no forums for engaging in an institutionalised dialogue with the 39 MPs from the Northeast. These institutional gridlocks have permitted few opportunity structures for learning and adaptation among key actors. If effectively shepherded, this could prove to be a valuable policy tool to bring a people-centric sensibility to questions of benefit sharing, risk allocation and trade-offs.

Governing regional public goods

Given these multiple ripple effects, resource management issues rightly need to be conceived as regional public goods that require transnational frames as against the bilateral frames (Kurian 2014). But are we making that causal link yet? And if so, how are we framing it? The increased pressure on resources, erosion concerns and water diversion plans raise livelihood questions for river-dependent communities both within and beyond borders. The livelihood chances and well-being of millions of largely rural riparian communities also need to be read together in the same frame as questions of accountability and representation. It is evident synergies with other public goods such as peace and stability also need to be acknowledged. It is this intrinsic nature that makes transboundary water management a 'means type public good' given that it can be an important means to safeguard other critical regional public goods such as peace, stability and biodiversity conservation (Andersen and Lindnaes 2007). Can we frame some of these questions in ways that can create institutional entry points for a whole set of missing issues that currently are invisible to the mainstream policy and research gaze in India and China? India and China's willingness to begin a sub-regional conversation on regional public goods could pave the way to designing norms of benefit sharing, negotiating trade-offs and allocating risks and burdens on collective goods and bads in the region.

Acknowledgements

The author wishes to thank the anonymous referee for comments and for comments received during the *National Workshop on Water Conflicts in the Northeast: Issues, Cases and Way Forward*, organised by the Forum for Policy Dialogue on Water Conflicts in India, Aaranyak and Action Aid, Guwahati, 10–11 December 2010.

Notes

1 China's western region has a high degree of ethnic diversity with ethnic minorities constituting 8.49 per cent of the population, making up 113.8 million and inhabiting 64 per cent of the country's total area, mostly along international borders.
2 Official admission of the less-than-expected benefits from the Three Gorges is revealing. In recent years, Chinese officials have downgraded the dam's flood-control capacity from 'able to withstand the worst flood in 10,000 years' in 2003, to 1,000 years in 2007, and to 100 years in 2008.
3 The Western Development Strategy was introduced during the Jiang Zemin-Zhu Rongji administration in 1999 and covers Chongqing Municipality, Sichuan, Guizhou, Yunnan, Shaanxi, Guansu and Qinghai provinces, Tibet, Ningxia Hui, Xinjiang Uygur, Inner Mongolia and Guangxi Zhuang autonomous regions. (For details, see Lan Xinzhen, 'A New Direction for West China', *Beijing Review*, 21 January 2010.)
4 For more on 'transitional theory'; K. Erhrhardt-Martinez, E. M. Crenshaw, and J. C. Jenkins. 2002. Deforestation and the Environmental Kuznet's curve: A cross-national investigation of intervening mechanisms. *Social Science Quarterly*, 83(1): 226–243.
5 Each MOU is for a duration of five years and has to be subsequently renegotiated for continued release of data.

6 The Indian side is led by the Commissioner, Ministry of Water Resources, and the Chinese side is led by Director, International Economic and Technical Cooperation and Exchange Centre, Ministry of Water Resources.

References

Andersen, E. A. and B. Lindnaes (eds.). 2007. *Towards New Global Strategies: Public Goods and Human Rights*. Leiden: Martinus Nijhoff Publishers.

Gunn, A. M. 2008. *Encyclopedia of Disasters: Environmental Catastrophes and Human Tragedies*. Westport, CT: Greenwood Press.

Huang, X. *et al*. 2010. Environmental impact of mining activities on the surface water quality in Tibet: Gyama valley. *The Science of the Total Environment*, 408(19), 1 September: 4177–4184. doi:10.1016/j.scitotenv.2010.05.015.

Kerr, R. and R. Stone. 2009. A human trigger for the great quake of Sichuan? *Science*, 323(5912), January: 322. doi: 10.1126/science.323.5912.322.

Kumar, V. 2013. Expert group calls for monitoring China's run-of-the-river projects. *The Hindu*, 10 April.

Kurian, N. 2004. Takes two to solve a water crisis. *Indian Express*, 17 August.

Kurian, N. 2009. 'A Glass Half Empty: Challenges to Subregional Water Dialogue'. In: Sharma, D. and L. Noronha (eds.). *Resource Security: The Governance Dimension*. The Energy and Resources Institute & Konrad Adenauer Stiftung. New Delhi: KAS Publication Series No. 27.

Kurian, N. 2014. *The India China Borderlands: Conversations Beyond the Centre*. New Delhi: Sage.

Magee, D. 2006. Powershed politics: Yunnan hydropower under great western development. *China Quarterly*, 185, March 2006: 23–41.

Mehdudia, S. 2011. We trust China's promise on dam: Manmohan. *The Hindu*, 5 August.

Ministry of External Affairs, Government of India. 2013. 'China Building Dam on Brahmaputra'. www.mea.gov.in/rajya-sabha.htm?dtl/22044/Q+514+CHINA+BUILDING+DAM+ON+BRAHMAPUTRA.

Ministry of External Affairs, Government of India. 2014. 'Water Sharing Relations with China'. http://meaindia.nic.in/mystart.php?id=100518234.

Overseas Development Institute. 2000. *Transboundary Water Management as an International Public Good*. Stockholm: Department of International Development Cooperation.

Pande, A. 2014. PM Narendra Modi should resurrect inter-state council to firm up his federalist intentions. *Economic Times*, 2 September.

Press Information Bureau. 2014. Government of India. *Ministry of External Affairs*, 13 August.

Press Trust of India (PTI). 2015. 'Centre not serious on issue of dams on Brahmaputra by China, says Assam CM Tarun Gogoi', 12 August. http://www.news18.com/news/politics/centre-not-serious-on-issue-of-dams-on-brahmaputra-by-china-says-assam-cm-tarun-gogoi-1038797.html (Accessed on 8 May 2017).

Ramachandran, S. 2008. Greater China: India quakes over China's water plan. *Asia Times Online*, 9 December.

Simons, C. Solving the entire Chinese water crisis. *The Atlanta Journal-Constitution*. www.progress.org/2006/water29.htm.

ThreeGorgesProbe.org. September 2006. http://www.tew.org/editorial-oped/trin-gyi-pho-nya/1106.html (Accessed on 8 May 2017).

United Nations Development Programme (UNDP), Human Development Report. 2006. *Beyond Scarcity: Power, Poverty and the Global Water Crisis*. New York: UNDP.

Valdiya, K. S. 1992. Must we have high dams in the geo-dynamically active Himalayan domain? *Current Science* (India), 25 September.

Wang, Q. 2010. 'China pledges water will still flow'. www.chinadaily.com.cn/china/2010-11/19/content_11574103.htm.

Milton Keynes UK
Ingram Content Group UK Ltd.
UKHW020000071024
449327UK00031B/2593